GEORGIA 1ST GRADE CCGPS MATHEMATICS REVIEW

ERICA DAY

COLLEEN PINTOZZI

MARY REAGAN

American Book Company

P. O. Box 2638

Woodstock, Georgia 30188-1383

Toll Free 1 (888) 264-5877 Phone (770) 928-2834

Toll Free Fax 1 (866) 827-3240

WEB SITE: www.americanbookcompany.com

Acknowledgements

In preparing this book, we would like to acknowledge Mary Reagan and Eric Field for their contributions developing graphics and Camille Woodhouse for her contributions in editing this book. We would also like to thank our many students whose needs and questions inspired us to write this text.

Copyright © 2010, 2009
by American Book Company
P.O. Box 2638
Woodstock, GA 30188-1383

ALL RIGHTS RESERVED

The text of this publication, or any part thereof, may not be reproduced or transmitted in any form or by any means, electronic or mechanical, including photocopying, recording, storage in an information retrieval system, or otherwise, without the prior permission of the publisher.

Printed in the United States of America
05/11   06/10   08/09

Contents

Acknowledgements		ii
Preface		viii
Diagnostic Test		1
	Part 1	1
	Part 2	10
1 Whole Numbers		**21**
1.1	Whole Numbers	21
1.2	Number Words	24
1.3	Modeling Numbers	25
1.4	Counting Numbers	26
1.5	Comparing Numbers	29
1.6	Ordering Numbers on a Number Line	33
1.7	Counting on a Number Line	35
	Chapter 1 Review	37
	Chapter 1 Test	40
2 Place Value		**43**
2.1	Tens and Ones	43
2.2	Modeling Tens and Ones	46
2.3	More Practice	48
2.4	Estimate to the Nearest Ten	49
2.5	Estimate to the Nearest Ten on a Number Line	50
2.6	Estimate to the Nearest Ten Using a Hundreds Chart	51
	Chapter 2 Review	52
	Chapter 2 Test	54
3 Addition		**57**
3.1	Skip Counting Forward	57
3.2	Addition Sentences	59
3.3	Adding with Zero	61

	3.4	Compose Numbers	62
	3.5	Counting On	64
	3.6	Doubles	66
	3.7	Making Tens	68
	3.8	Making More Tens and Ones	69
	3.9	Adding in Any Order	70
	3.10	Adding Across and Down	71
	3.11	Adding 3 Numbers	72
	3.12	More Practice	73
	3.13	Word Problems	75
		Chapter 3 Review	76
		Chapter 3 Test	78
4	**Subtraction**		**80**
	4.1	Skip Counting Backwards	80
	4.2	Subtraction Sentences	82
	4.3	Subtracting All or Zero	84
	4.4	Decompose Numbers Up to 10	85
	4.5	Counting Back	87
	4.6	Fact Families	89
	4.7	Subtracting Across and Down	90
	4.8	Comparing	91
	4.9	More Practice	93
	4.10	Word Problems	94
		Chapter 4 Review	95
		Chapter 4 Test	97
5	**Number Patterns**		**99**
	5.1	Patterns on a Hundred Chart	99
	5.2	Even and Odd Patterns	101
	5.3	Using Skip Counting for Number Patterns	104
	5.4	Number Patterns	107
	5.5	Use and Share Objects	108
		Chapter 5 Review	110

Contents

        Chapter 5 Test   112

6 Fractions   **114**
- 6.1 Equal Parts   114
- 6.2 Halves   116
- 6.3 Fourths   117
- 6.4 More Practice   118
-     Chapter 6 Review   119
-     Chapter 6 Test   120

7 Money   **122**
- 7.1 Pennies and Nickels   122
- 7.2 Dimes and Quarters   124
- 7.3 Counting Coins   126
- 7.4 Fair Trades   128
- 7.5 Making Purchases   129
- 7.6 Dollar Bills   130
- 7.7 Fair Trades Using Bills   132
- 7.8 Making Purchases Using Bills   133
-     Chapter 7 Review   134
-     Chapter 7 Test   136

8 Measurement   **138**
- 8.1 Compare Length   138
- 8.2 Measure Length Using Nonstandard Units   140
- 8.3 Measure Length with a Tool   141
- 8.4 Inches and Centimeters   142
- 8.5 Compare Weight   143
- 8.6 Measure Weight Using Nonstandard Units   144
- 8.7 Compare Capacity   146
- 8.8 Measure Capacity Using Nonstandard Units   147
- 8.9 Measure Capacity with a Tool   148
-     Chapter 8 Review   149
-     Chapter 8 Test   151

Contents

9 Time — **154**
- 9.1 Tell Time to the Nearest Hour — 154
- 9.2 Tell Time to the Nearest Half Hour — 156
- 9.3 Time for More Practice — 157
- 9.4 Sequence of Events — 158
- 9.5 Duration of Events — 159
- 9.6 Calendar — 160
- Chapter 9 Review — 162
- Chapter 9 Test — 165

10 Geometry — **168**
- 10.1 Plane Figures — 168
- 10.2 Solid Figures — 170
- 10.3 Naming Faces of Solid Figures — 172
- 10.4 Compare Shapes — 174
- 10.5 Create Pictures Using Shapes — 175
- 10.6 Overlapping Shapes — 177
- 10.7 Position — 178
- 10.8 Direction — 180
- Chapter 10 Review — 181
- Chapter 10 Test — 183

11 Graphs — **186**
- 11.1 Tables — 186
- 11.2 Tally Charts — 188
- 11.3 Making Tables and Tally Charts — 190
- 11.4 Picture Graphs — 191
- 11.5 Bar Graphs — 193
- 11.6 Making Bar Graphs — 194
- Chapter 11 Review — 195
- Chapter 11 Test — 197

12 Addition and Subtraction with 2-Digit Numbers — **199**
- 12.1 Add Tens — 199

Contents

12.2	Add Tens and Ones	200
12.3	Add 2-Digit Numbers	201
12.4	Subtract Tens	202
12.5	Subtract Tens and Ones	203
12.6	Subtract 2-Digit Numbers	204
12.7	Mixed Practice	205
12.8	Adding Money	206
12.9	Subtracting Money	207
12.10	Choose the Operation	208
12.11	Word Problems	209
	Chapter 12 Review	210
	Chapter 12 Test	211

Practice Test 1 — 213
    Part 1 — 213
    Part 2 — 222

Practice Test 2 — 231
    Part 1 — 231
    Part 2 — 240

Preface

Georgia 1st Grade CCGPS Mathematics Review will help students who are learning or reviewing the Georgia Performance Standards for the first grade. The materials in this book are based on the CCGPS as published by the Georgia Department of Education. This book is written to the grade 1 level.

This book contains several sections:

1) General information about the book itself
2) A diagnostic test
3) An evaluation chart
4) Twelve chapters that teach the concepts and skills needed for test readiness
5) Two practice tests

The complete list of standards is located in the Answer Key. Each question in the Diagnostic and Practice Tests is referenced to the standard, as is the beginning of each chapter.

ABOUT THE AUTHORS

Erica Day has a Bachelor of Science Degree in Mathematics and is working on a Master of Science Degree in Mathematics. She graduated with high honors from Kennesaw State University in Kennesaw, Georgia. She has also tutored all levels of mathematics, ranging from high school algebra and geometry to university-level statistics, calculus, and linear algebra.

Colleen Pintozzi has taught mathematics at the middle school, junior high, senior high, and adult level for 22 years. She holds a B.S. degree from Wright State University in Dayton, Ohio and has done graduate work at Wright State University, Duke University, and the University of North Carolina at Chapel Hill.

Diagnostic Test

Part 1

Read the directions and circle the right answer.

1. What is nine in number form?

   A 5
   B 7
   C 9

   M1N1a

2. Count how many.

   A 4
   B 3
   C 2

   M1N1b

3. Which number sentence is correct?

   A 5 < 3
   B 3 < 5
   C 3 > 5

   M1N1c

4. Which is a fair trade?

   A

   B

   C

   M1N1e

5. What is the value of the bill below?

A $1.00
B $5.00
C $50.00

M1N1f

6. Look at the numbers chart below. Which number is 73 closest to?

51	52	53	54	55	56	57	58	59	60
61	62	63	64	65	66	67	68	69	70
71	72	73	74	75	76	77	78	79	80

A 60
B 70
C 80

M1N2a

7. How many groups of tens?

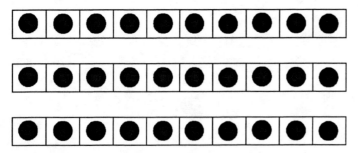

A 4 groups of tens
B 3 groups of tens
C 2 groups of tens

M1N2b

8. How many tens and ones does 18 have?

   A
Tens	Ones
8	1

   B
Tens	Ones
1	8

   C
Tens	Ones
9	9

9. What is one more than 14?

   A 15
   B 16
   C 17

10. Which number sentence is equal to 9?

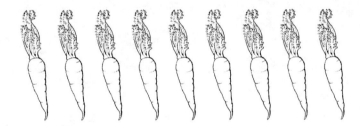

   A $3 + 9 = 9$
   B $4 + 6 = 9$
   C $3 + 6 = 9$

11. You have 3 eggs. You count on 2 more. How many eggs do you have now?

   A 3
   B 4
   C 5

12. Add. $\begin{array}{r}12\\+34\\\hline\end{array}$

   A 46
   B 36
   C 16

13. Five geese flew to the lake. Three of the geese stayed at the lake. How many geese flew away from the lake?

   A 5
   B 3
   C 2

14. Taye ate 4 cookies. Then he ate 2 more. How many cookies did Taye eat in all?

   A 6
   B 5
   C 4

15. Amy's mouse had 6 baby mice.
    She gave half the babies to Kali and half to Jan.
    How many mice will Kali and Jan each have?

   A 6
   B 3
   C 2

16. Complete the sentence:
    The glass is _____ .

   A quarter full
   B half full
   C full

17. Which worm is the shortest?

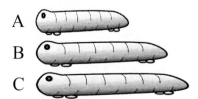

18. About how many paper clips long is the crayon?

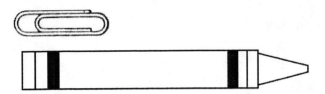

    A 1 paper clip long

    B 2 paper clips long

    C 3 paper clips long

M1M1b

19. About how many thumb widths is the fish?
    (Use the thumbs width in the drawing, not your own thumb.)

    A 3
    B 4
    C 5

M1M1c

20. What time is it to the nearest hour?

    A 2 : 00
    B 3 : 00
    C 11 : 00

M1M2a

21. Which takes the longest to do?

   A Saying the alphabet song.
   B Eating your lunch.
   C Sleeping the whole night.

22. What is the name of this shape?

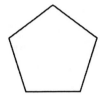

   A square
   B pentagon
   C hexagon

23. How many sides does a rectangle have?

   A 3
   B 4
   C 5

24. What is the name of the object below?

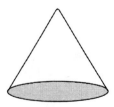

   A cube
   B cylinder
   C cone

25. Which solid is this glass most like?

   A cube
   B cylinder
   C cone

26. Which of these shapes has 3 sides and 3 corners?

   Triangle **Square** **Hexagon**

   A triangle
   B square
   C hexagon

27. Where is the ball?

   A The ball is under the elephant.
   B The ball is in back of the elephant.
   C The ball is above the elephant.

Use this chart for the next 3 problems.

Favorite Pies

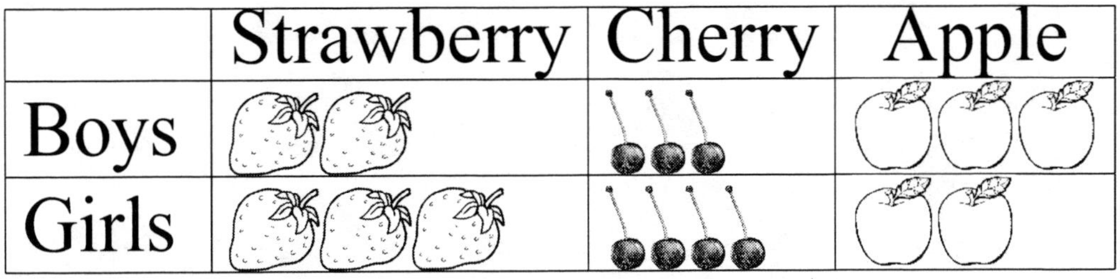

	Strawberry	Cherry	Apple
Boys	🍓🍓	🍒🍒🍒	🍎🍎🍎
Girls	🍓🍓🍓	🍒🍒🍒🍒	🍎🍎

Key: Each piece of fruit equals one vote

28. How many boys like cherry pie best?

   A 2
   B 3
   C 5

29. How many of the girls like strawberry pie best?

   A 3
   B 4
   C 6

30. Which tally chart shows the pies the boys like best?

Chart 1	Count
Strawberry	/ /
Cherry	/ / /
Apple	/ / /

Chart 2	Count
Strawberry	/ / /
Cherry	/ /
Apple	/ / / /

Chart 3	Count
Strawberry	/ /
Cherry	/ /
Apple	/ / /

   A Chart 1
   B Chart 2
   C Chart 3

Part 2

31. ★★★★★★★★★★ + ★★★★★★★★★★ + ★★★★ = _____

   A 20
   B 24
   C 34

   M1N1a

32. Count how many.

   A 4
   B 5
   C 6

   M1N1b

33. Which number sentence is correct?

   A 8 > 4
   B 8 < 4
   C 4 > 8

   M1N1c

34. Look at the number line. What number is missing?

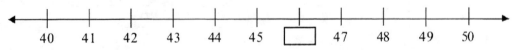

   A 44
   B 45
   C 46

   M1N1d

35. Juan has 89 ¢. The toy he wants costs 79 ¢.
    Does Juan have enough money? Will he have money leftover?

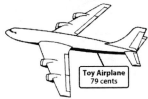

   A Yes, and he will have 10 ¢ leftover.
   B Yes, and he will have 5 ¢ leftover.
   C No, he doesn't have enough money.

M1N1e

36. Look at the number below.
    Which number is it closest to?

    67

    A 50
    B 60
    C 70

M1N2a

37. How many tens and ones does 37 have?

   A | Tens | Ones |
      |------|------|
      | 3 | 7 |

   B | Tens | Ones |
      |------|------|
      | 7 | 3 |

   C | Tens | Ones |
      |------|------|
      | 37 | 0 |

M1N2c

38. Skip count by 2's. What is the missing number?

    2, 4, ___, 8, 10

    A 4
    B 6
    C 8

M1N3b

39. Which number sentence is equal to 7?

    A $3 + 5 = 7$
    B $4 + 4 = 7$
    C $3 + 4 = 7$

M1N3c

40. Count how many ants are in the 1st set.
    Count how many ants are in the 2nd set.
    How many more ants are in Set 1?

A 2
B 3
C 5

M1N3d

41. Start at 55. Count backwards 2 numbers.
    What is the new number?

    A 55
    B 54
    C 53

M1N3e

42. Which sentence is true?

    A $5 + 3 = 3 + 5$
    B $5 + 3 = 53$
    C $5 + 3 = 2$

M1N3f

43. $\begin{array}{r}78\\-22\\\hline\end{array}$

   A 100
   B 66
   C 56

44. Mary had 4 balloons.
    She gave 2 balloons to her little sister.
    How many balloons does Mary have left?

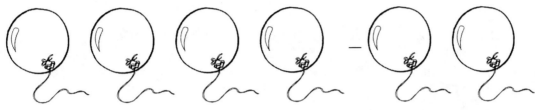

   A 6
   B 4
   C 2

45. Jake found 6 frogs.
    He shared them equally with 2 friends, Mark and Peter.
    How many frogs will Jake, Mark, and Peter each have?

   A 2
   B 3
   C 6

46. Is the number 17 even or odd?

   A even
   B odd
   C both

47. How many parts does the circle have?

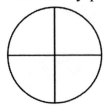

- A 2
- B 3
- C 4

48. Which weighs the most? A radish, a watermelon, or a pear?

- A radish
- B watermelon
- C pear

49. About how many paper clips long is the feather?

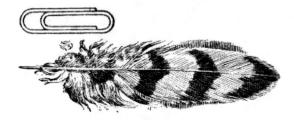

- A 1 paper clip long
- B 2 paper clips long
- C 3 paper clips long

50. The first three months of the year are January, February, and March. What are the next three months of the year?

   A April, May, and July

   B April, June, and July

   C April, May, and June

M1M2b

51. Which takes the shortest amount of time to do?

   A Count to 20.

   B Eat your dinner.

   C Go to school for a whole day.

M1M2c

52. Which shape is the checker board most like?

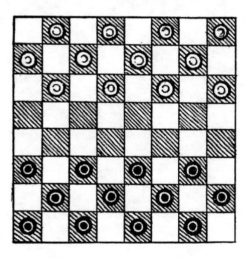

   A square

   B rectangle

   C triangle

M1G1a

53. How many sides does a triangle have?

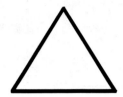

   A 3
   B 4
   C 5

54. What is the name of the object below?

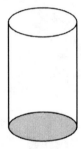

   A cube
   B cylinder
   C cone

55. Which solid is the box most like?

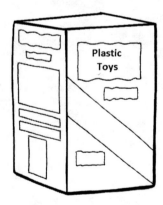

   A cube
   B rectangular prism
   C cone

56. Which of these shapes has 6 sides and 6 corners?

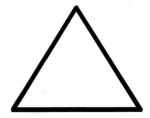 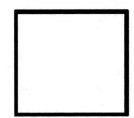 

   A triangle

   B square

   C hexagon

57. Where is the ball?

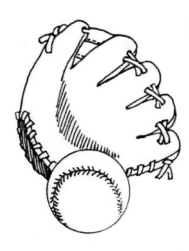

   A The ball is under the glove.

   B The ball is in front of the glove.

   C The ball behind the glove.

Use this chart for the next 3 problems.

58. How many quarters does Kiya have?

    A 3
    B 2
    C 5

    M1D1a

59. How many nickels does John have?

    A 4
    B 3
    C 2

    M1D1a

60. Which tally chart shows the coins that Kiya has?

Chart 1	Count
Quarters	/ / /
Dimes	/ /
Nickels	/ / /

Chart 2	Count
Quarters	/ / /
Dimes	/ /
Nickels	/ /

Chart 3	Count
Quarters	/ /
Dimes	/ / /
Nickels	/ /

A Chart 1

B Chart 2

C Chart 3

M1D1b

Evaluation Chart for the Diagnostic Mathematics Test

Directions: On the following chart, circle the question numbers that you answered incorrectly. Then turn to the chapter(s), read, and complete the problems. Review the other chapters as needed. Finally, complete the *Georgia 1st Grade CCGPS Mathematics Review* Practice Tests to further review.

		Questions Part 1	Questions Part 2	Pages
Chapter 1:	Whole Numbers	1, 2, 3	31, 32, 33, 34	21–42
Chapter 2:	Place Value	6, 7, 8	36, 37	43–56
Chapter 3:	Addition	10, 11, 14	38, 39, 42	57–79
Chapter 4:	Subtraction	13	40, 41, 44	80–98
Chapter 5:	Number Patterns	9, 15	45, 46	99–113
Chapter 6:	Fractions	16	47	114–121
Chapter 7:	Money	4, 5	35	122–137
Chapter 8:	Measurement	17, 18, 19	48, 49	138–153
Chapter 9:	Time	20, 21	50, 51	154–167
Chapter 10	Geometry	22, 23, 24 25, 26, 27	52, 53, 54 55, 56, 57	168–185
Chapter 11	Graphs	28, 29, 30	58, 59, 60	186–198
Chapter 12	Addition and Subtraction with 2-Digit Numbers	12	43	199–212

Chapter 1
Whole Numbers

This chapter covers the following Georgia Performance Standards:

| M1N | Number and Operations | M1N1a, b, c, d |

Whole Numbers

A number tells how many.

1 one ☺
2 two ☺ ☺
3 three ☺ ☺ ☺
4 four ☺ ☺ ☺ ☺
5 five ☺ ☺ ☺ ☺ ☺
6 six ☺ ☺ ☺ ☺ ☺ ☺
7 seven ☺ ☺ ☺ ☺ ☺ ☺ ☺
8 eight ☺ ☺ ☺ ☺ ☺ ☺ ☺ ☺
9 nine ☺ ☺ ☺ ☺ ☺ ☺ ☺ ☺ ☺
10 ten ☺ ☺ ☺ ☺ ☺ ☺ ☺ ☺ ☺ ☺

Chapter 1   Whole Numbers

Count. Write the number.

1. _____
   _ _ _ _ _
   _____
   three

2. _____
   _ _ _ _ _
   _____
   one

3. _____
   _ _ _ _ _
   _____
   five

4. _____
   _ _ _ _ _
   _____
   six

5. _____
   _ _ _ _ _
   _____
   four

Whole Numbers

6.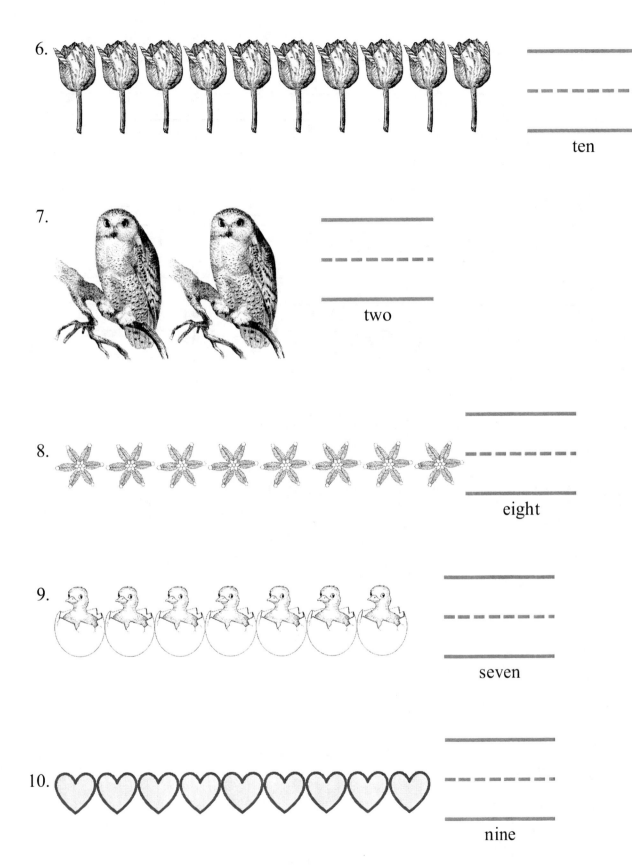

ten

7.

two

8.

eight

9.

seven

10.

nine

Chapter 1   Whole Numbers

Use the hundreds chart to count to 100.

Hundreds Chart

1	2	3	4	5	6	7	8	9	10
11	12	13	14	15	16	17	18	19	20
21	22	23	24	25	26	27	28	29	30
31	32	33	34	35	36	37	38	39	40
41	42	43	44	45	46	47	48	49	50
51	52	53	54	55	56	57	58	59	60
61	62	63	64	65	66	67	68	69	70
71	72	73	74	75	76	77	78	79	80
81	82	83	84	85	86	87	88	89	90
91	92	93	94	95	96	97	98	99	100

Number Words

EXAMPLE: The number **1** is written **one**.

EXAMPLE: The number **84** is written **eighty-four**.

Read the sentence. Choose the number from the chart.

96	2	24	9	12	38	64	73

Fill in the blanks.

1. The number two is _____ .

2. The number seventy-three is _____ .

3. The number nine is _____ .

4. The number twenty-four is _____ .

5. The number twelve is _____ .

6. The number thirty-eight is _____ .

7. The number ninety-six is _____ .

8. The number sixty-four is _____ .

Modeling Numbers

First, count and circle the number that tells how many in all.

1. Next, color 3 acorns brown. Then color 4 acorns green.

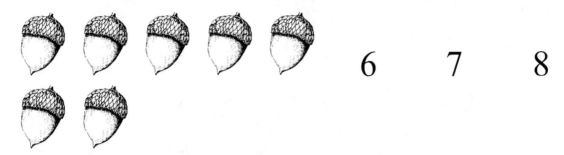   6   7   8

2. Color 2 eggs blue. Then color 7 eggs yellow.

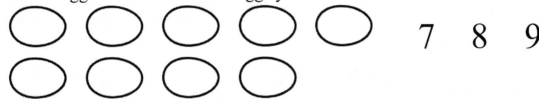   7   8   9

3. Color 3 diamonds green. Then color 3 diamonds purple.

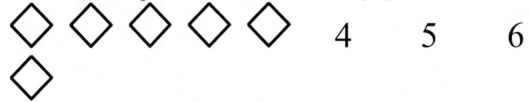   4   5   6

4. Color 4 hearts red. Then color 4 hearts purple.

   6   7   8

5. Color 6 suns yellow. Then color 4 suns orange.

   8   9   10

Chapter 1   Whole Numbers

Counting Numbers

Count to find out how many.

EXAMPLE:    3

EXAMPLE:    5

Count. Write the number.

1.    _____

2.    _____

3.    _____

4.    _____

5.    _____

Counting Numbers

Match groups and numbers. Draw a line to the match.

1.   10

2.   3

3.   5

4.   7

5.   6

Chapter 1   Whole Numbers

Use the picture. Write how many.

1. _____

2. _____

3. _____

****

4. _____

****

5. _____

Comparing Numbers

Count how many. Write the number.
Circle the number that has more.

1.

2.

3.

4.

Chapter 1   Whole Numbers

Count how many. Write the number.
Circle the number that has <u>less</u>.

1.

2.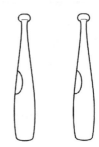

3.

4.

Comparing Numbers

Greater Than

The sign for greater than is >.

EXAMPLE: **57** is **greater than** **54**.
57 and 54 have the same number of tens, but 57 has more ones than 54.

57 > 54

Read the two numbers. Fill in the number sentences.

1. 22   20

   _____ is greater than _____ .

   _____ > _____ .

2. 37   39

   _____ is greater than _____ .

   _____ > _____ .

3. 88   84

   _____ is greater than _____ .

   _____ > _____ .

4. 67   63

   _____ is greater than _____ .

   _____ > _____ .

5. 43   47

   _____ is greater than _____ .

   _____ > _____ .

Chapter 1   Whole Numbers

Less Than

The sign for <u>less than</u> is <.

EXAMPLE: <u>23</u> is **less than** <u>28</u>.
23 and 28 have the same number of tens, but 23 has less ones than 28.

<u>23</u> < <u>28</u>

Read the two numbers. Fill in the number sentences.

1. 71   74

    _____ is less than _____ .

    _____ < _____ .

2. 92   99

    _____ is less than _____ .

    _____ < _____ .

3. 47   44

    _____ is less than _____ .

    _____ < _____ .

4. 37   32

    _____ is less than _____ .

    _____ < _____ .

5. 11   16

    _____ is less than _____ .

    _____ < _____ .

Ordering Numbers on a Number Line

The numbers are in order.

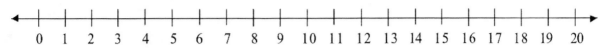

EXAMPLE: 4 comes before 5 on the number line.

EXAMPLE: 15 comes before 16 on the number line.

Fill in the blank. Write the number that comes just <u>before</u>.

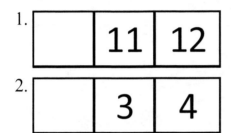
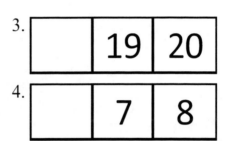

Fill in the blank. Write the number that comes just <u>after</u>.

5. | 8 | 9 | □ |
6. | 12 | 13 | □ |

7. | 17 | 18 | □ |
8. | 2 | 3 | □ |

Fill in the blank. Write the number that comes <u>between</u>.

9. | 10 | □ | 12 |
10. | 6 | □ | 8 |

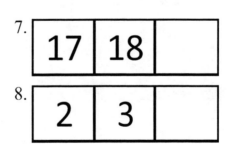

Chapter 1   Whole Numbers

Find the picture. Then find the number below the picture.

EXAMPLE: Find the 🐜. The 🐜 is located at what point?

The 🐜 is located at point 3.

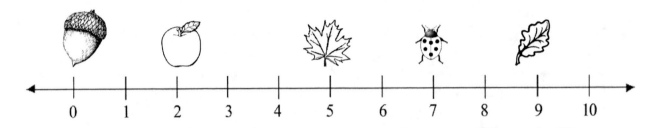

1. The 🍁 is located at what point? _____

2. The 🌰 is located at what point? _____

3. The 🍃 is located at what point? _____

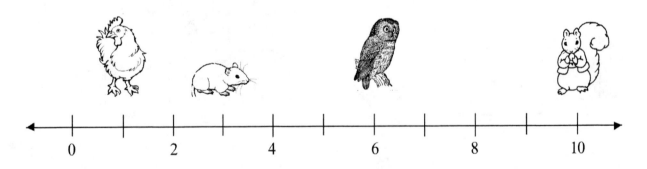

4. The 🐿 is located at what point? _____

5. The 🐭 is located at what point? _____

6. The 🦉 is located at what point? _____

Counting on a Number Line

Count **forward**. Write the numbers.
Use the number line to help you.

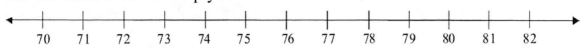

1. Count forward from 72.

    72, _____, _____, _____, _____

2. Count forward from 78.

    78, _____, _____, _____, _____

3. Count forward from 71.

    71, _____, _____, _____, _____

4. Count forward from 79.

    79, _____, _____, _____, _____

5. Count forward from 77.

    77, _____, _____, _____, _____

6. Count forward from 73.

    73, _____, _____, _____, _____

Chapter 1   Whole Numbers

Count **backward**. Write the numbers.
Use the number line to help you.

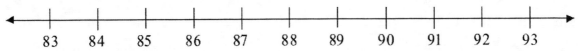

1. Count backward from 88.

   88, _____, _____, _____

2. Count backward from 86.

   86, _____, _____, _____

3. Count backward from 90.

   90, _____, _____, _____

4. Count backward from 85.

   85, _____, _____, _____

5. Count backward from 89.

   89, _____, _____, _____

6. Count backward from 92.

   92, _____, _____, _____

Chapter 1 Review

Count. Write the number.

1.  _____

2.  _____

3.  _____

Fill in the blanks.

4. The number two is _____ .

5. The number seventeen is _____ .

6. The number sixty-three is _____ .

Count. Circle the number that tells how many.

7. ★ ★ ★ ★ ★
   ★ ★   6   7   8

Chapter 1   Whole Numbers

Count. Write the number.

8.

9.

Use the picture. Write how many.

10.

**Count how many. Write the number.
Circle the number that has more.**

11.

12.

Chapter 1 Review

**Count how many. Write the number.
Circle the number that has less.**

13.  _____

14.  _____

Use the number line. Write the number that is missing.

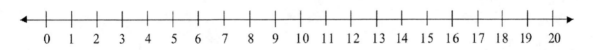

15. | 16 | | 18 |

16. | 4 | 5 | |

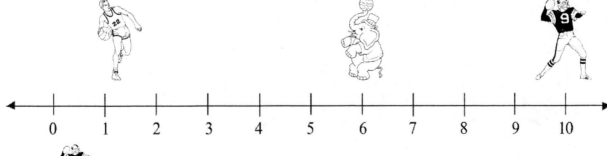

17. The is located at what point? _____

18. Count **forward** from 17, ____, ____, ____

19. Count **backwards** from 64, ____, ____, ____

Chapter 1 Test

Read the directions and circle the right answer.

1. Count the balls.

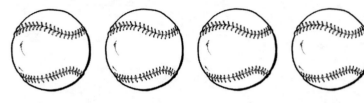

   Circle the number of balls.

   A 2

   B 3

   C 4

2. Fill in the blank:

   The number eight is _____ .

   A 7

   B 8

   C 9

3. Count.

   Circle the number of balls.

   A 5

   B 6

   C 7

Chapter 1 Test

4. Use the picture.

Circle the number of fish.

A 3

B 4

C 5

5. Count the 2 groups.

What is the number of the group that has more?

A 2

B 4

C 5

Chapter 1   Whole Numbers

6. Count the 2 groups.

What is the number of the group that has less?

A 2

B 3

C 5

7. What is the missing number?

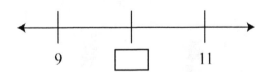

A 10

B 12

C 13

Chapter 2
Place Value

This chapter covers the following Georgia Performance Standards:

| M1N | Number and Operations | M1N2a, b, c |

Tens and Ones

Groups of Tens.

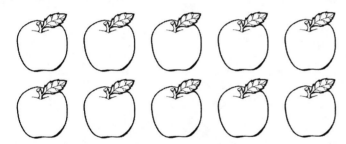

1 group of ten = 10.

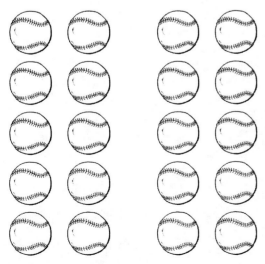

2 groups of ten = 20.

Chapter 2   Place Value

Count the groups of tens. Write the number.

1.  ____ groups of tens = ____ .

2.  ____ groups of tens = ____ .

3. ____ groups of tens = ____ .

4. ____ groups of tens = ____ .

Tens and Ones

Groups of Tens and Ones.

EXAMPLE: 34 = **3** tens and **4** ones.
34 has 3 groups of ten and 4 ones.
34 ones = 3 tens and 4 ones.

Tens	Ones
3	4

Read each number.
Write how many tens and ones.

1. **56**   ___ tens ___ ones

2. **17**   ___ tens ___ ones

3. **39**   ___ tens ___ ones

4. **72**   ___ tens ___ ones

5. **44**   ___ tens ___ ones

6. **85**   ___ tens ___ ones

Chapter 2 Place Value

Modeling Tens and Ones

EXAMPLE: 26

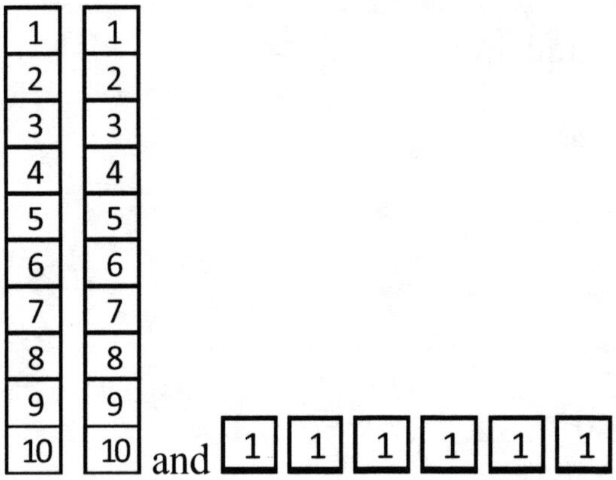

2 TENS AND 6 ONES

Count the number of tens and ones. Write the number.

1.

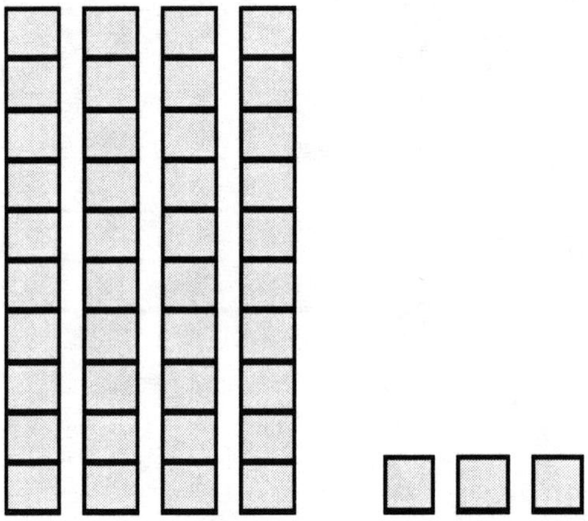

Number: ____

____ tens   ____ ones

2.

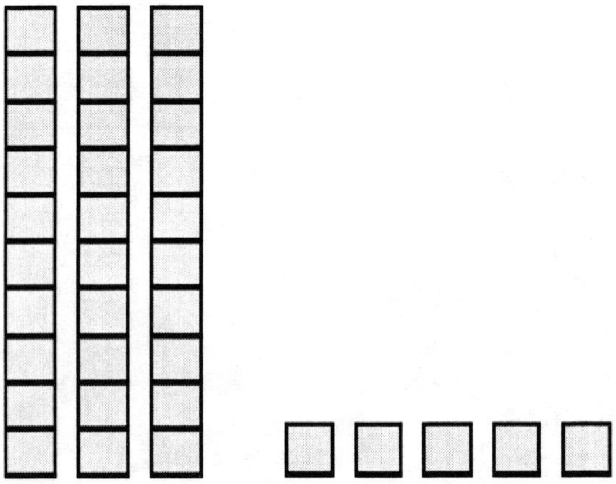

Number: ____

____ tens   ____ ones

Modeling Tens and Ones

Count the number of tens and ones. Write the number.

1. Color 31 squares red and 21 squares yellow.

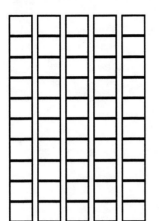

Number: ____

____ tens   ____ ones

2. Color 10 squares green and 4 squares blue.

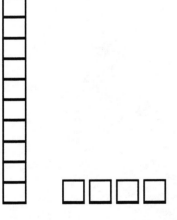

Number: ____

____ tens   ____ ones

3. Color 20 squares purple and 11 squares orange.

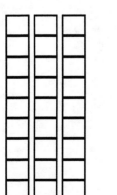

Number: ____

____ tens   ____ ones

Chapter 2   Place Value

More Practice

Write each number.

1. 3 tens   7 ones   _____

2. 6 tens   8 ones   _____

3. 9 tens   9 ones   _____

4. 2 tens   1 one   _____

5. 4 tens   6 ones   _____

6. 1 ten   2 ones   _____

7.
tens	ones
8	5

_____

8.
tens	ones
9	2

_____

9.
tens	ones
1	4

_____

10.
tens	ones
2	2

_____

11.
tens	ones
3	9

_____

12.
tens	ones
7	1

_____

Estimate to the Nearest Ten

You can **estimate** to find out how many.
An estimate is finding **about** how many.
It is not always the exact amount, but closest to the nearest ten.

EXAMPLE: About how many are in the row?
Circle your estimate.

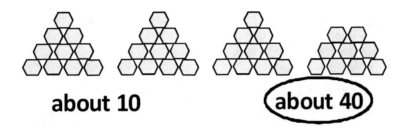

about 10   (about 40)

About how many are in the row?
Circle your estimate.

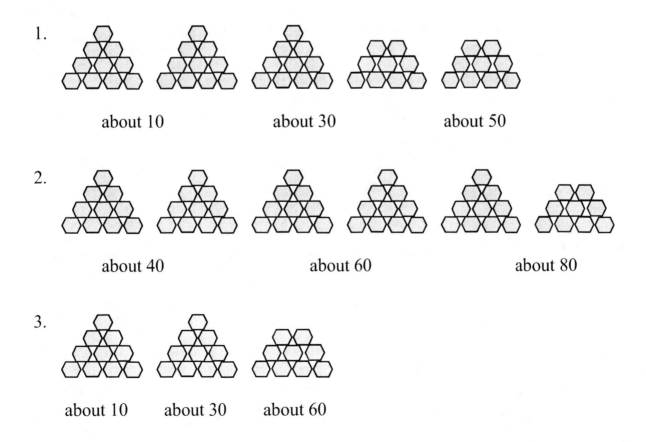

1. about 10   about 30   about 50

2. about 40   about 60   about 80

3. about 10   about 30   about 60

Chapter 2   Place Value

Estimate to the Nearest Ten on a Number Line

EXAMPLE: Look at the number line below.
Estimate 43 to the nearest ten.

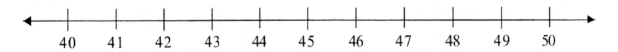

43 is nearer 40 than 50. So the answer is 40.

Use the number lines below.
Estimate each number to the nearest ten.
Circle your answer.

1.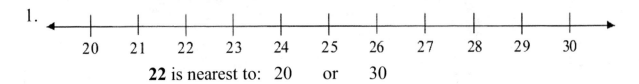
   22 is nearest to: 20   or   30

2.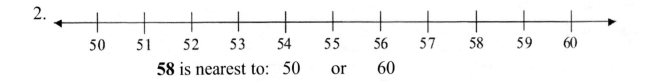
   58 is nearest to: 50   or   60

3.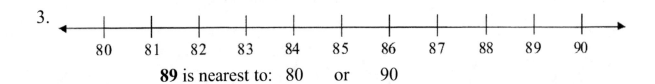
   89 is nearest to: 80   or   90

4.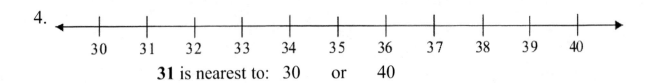
   31 is nearest to: 30   or   40

5.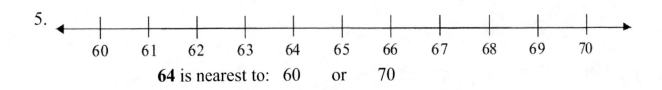
   64 is nearest to: 60   or   70

Estimate to the Nearest Ten Using a Hundreds Chart

1	2	3	4	5	6	7	8	9	10
11	12	13	14	15	16	17	18	19	20
21	22	23	24	25	26	27	28	29	30
31	32	33	34	35	36	37	38	39	40
41	42	43	44	45	46	47	48	49	50
51	52	53	54	55	56	57	58	59	60
61	62	63	64	65	66	67	68	69	70
71	72	73	74	75	76	77	78	79	80
81	82	83	84	85	86	87	88	89	90
91	92	93	94	95	96	97	98	99	100

Example: Tens ↑

Use the chart to estimate the nearest ten.
Circle the nearest ten.

1. **87** is nearest to: 70   80   90

2. **21** is nearest to: 20   30   40

3. **39** is nearest to: 30   40   50

4. **42** is nearest to: 30   40   50

5. **58** is nearest to: 40   50   60

6. **63** is nearest to: 50   60   70

7. **79** is nearest to: 70   80   90

Chapter 2   Place Value

Chapter 2 Review

Count the groups of tens. Write the number.

1. ♣ ♣ ♣ ♣ ♣
   ♣ ♣ ♣ ♣ ♣

   ♣ ♣ ♣ ♣ ♣
   ♣ ♣ ♣ ♣ ♣   ____ groups of tens = ____ .

2. ★ ★   ★ ★   ★ ★
   ★ ★   ★ ★   ★ ★
   ★ ★   ★ ★   ★ ★
   ★ ★   ★ ★   ★ ★
   ★ ★   ★ ★   ★ ★   ____ groups of tens = ____ .

Read each number.
Write how many tens and ones.

3. **28**   ____ tens ____ ones

4. **63**   ____ tens ____ ones

Count the number of tens and ones. Write the number.

5.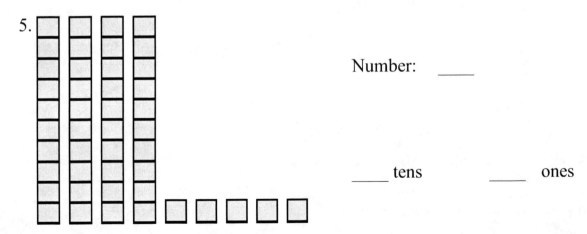

   Number: ____

   ____ tens   ____ ones

Chapter 2 Review

Write each number.

6. 8 tens and 3 ones _____

7. 5 tens and 4 ones _____

8.
tens	ones
1	7	_____

9.
tens	ones
4	1	_____

About how many are in the row?
Circle your estimate.

10.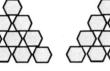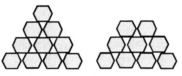

    about 10   about 40   about 60

11.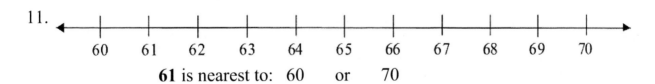

    61 is nearest to: 60   or   70

12.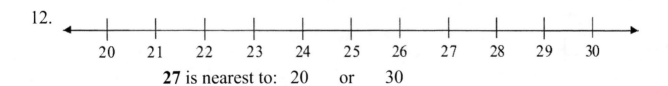

    27 is nearest to: 20   or   30

Circle the best estimate.

13. 31 is nearest to: 20   30   40

14. 88 is nearest to: 70   80   90

Chapter 2 Test

Read the directions and circle the right answer.

1. How many groups of tens are there?

   A 2 groups of ten = 20

   B 3 groups of ten = 30

   C 4 groups of ten = 40

2. Read the number: 91

   How many ones and tens are there?

   A 9 tens and 1 one

   B 1 ten and 9 ones

   C 91 tens and 91 ones

3. How many ones and tens are there?

   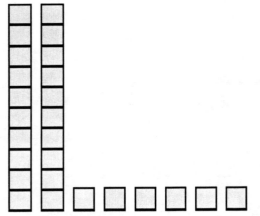

   A 2 tens and 4 ones

   B 2 tens and 5 ones

   C 2 tens and 6 ones

4. What do 7 tens and 5 ones equal?

   A 57

   B 75

   C 71051

5. What do 5 tens and 6 ones equal?

   A 56

   B 65

   C 51061

6. What number is:

tens	ones
6	3
   ?

   A 61031

   B 63

   C 36

7. What number is:

tens	ones
5	5
   ?

   A 55

   B 50

   C 5

8. Estimate how many are in the row.

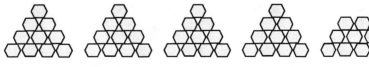

   A 60

   B 70

   C 80

Chapter 2 Place Value

9. Look at the number line below. What ten is 78 nearest to?

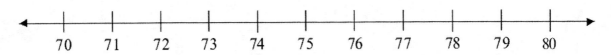

A 70

B 80

C 90

10. 51 is nearest to:

A 50

B 60

C 70

11. 92 is nearest to:

A 70

B 80

C 90

12. 11 is nearest to:

A 10

B 20

C 30

Chapter 3
Addition

This chapter covers the following Georgia Performance Standards:

| M1N | Number and Operations | M1N3b, c, e, f, h |

Skip Counting Forward

Skip counting is counting, but not using every number.
Sometimes we count by 2's, 5's, or 10's.

EXAMPLE: Skip count by twos.

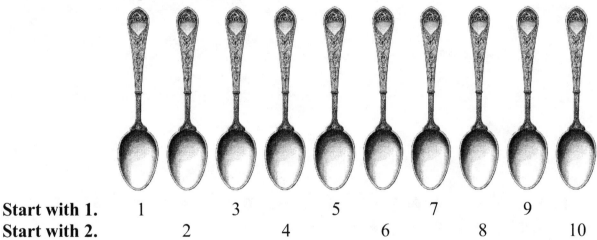

Start with 1. 1   3   5   7   9
Start with 2.   2   4   6   8   10

1. Skip count by twos. Write the numbers.

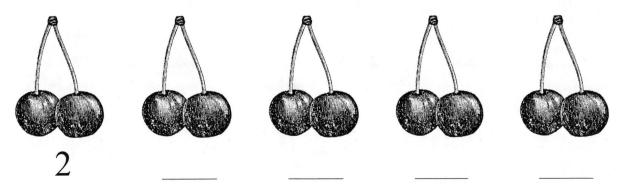

2   ___   ___   ___   ___

Chapter 3   Addition

EXAMPLE: Skip count by fives.

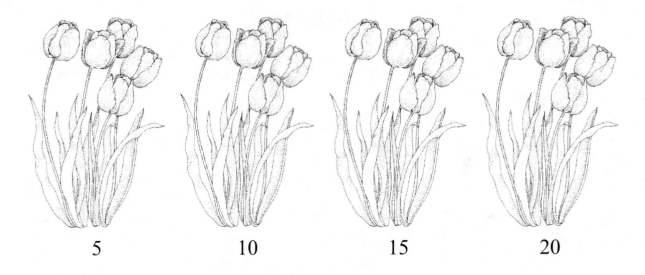

2. Skip count by fives. Write the numbers.

5   _____   _____   _____   _____

EXAMPLE: Skip count by tens.

3. Skip count by tens. Write the numbers.

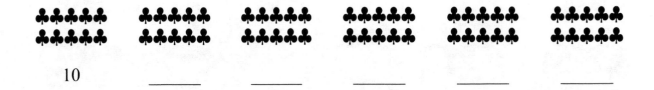

10   _____   _____   _____   _____   _____

Addition Sentences

2 and 3 is 5 in all.
2 plus 3 equals 5.
2 + 3 = 5
2 pumpkins + 3 pumpkins = 5 pumpkins

The + means plus.
The = means equal.
The total is the <u>sum</u>. 5 is the total. 5 is the <u>sum</u>.

Write in the + and the = in the ◯. The first one is done for you.

1. 1 + 2 = 3
2. 3 ◯ 2 ◯ 5
3. 1 ◯ 3 ◯ 4
4. 2 ◯ 2 ◯ 4

Add. Write the sum.

5. 1 + 4 = ___
6. 3 + 1 = ___
7. 1 + 1 = ___
8. 3 + 3 = ___

Chapter 3   Addition

**Use the pictures. Write in the missing + and = signs.
Add the numbers. Write in the sum.**

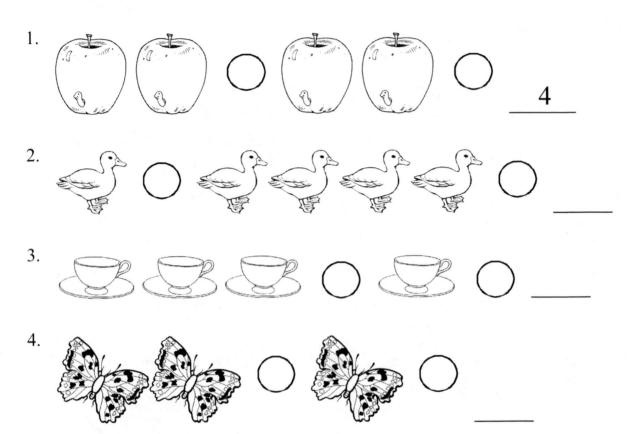

1. (apples) ○ (apples) ○ ___4___
2. (duck) ○ (ducks) ○ ___
3. (cups) ○ (cups) ○ ___
4. (butterflies) ○ (butterfly) ○ ___

**Use the pictures. Write the number sentence.
Add the numbers. Write in the sum.**

5. (4 acorns) + (2 acorns) = ___

Number sentence: _____

6. (1 bee) + (3 bees) = ___

Number sentence: _____

60

Adding with Zero

When you add zero to any number, the sum is the same as the number added to zero.

EXAMPLE: Add zero to 3.

$3 + 0 = 3$

EXAMPLE: Add zero to 7.
$7 + 0 = 7$

Add. Write the sum.

1. $1 + 0 =$ _____

2. $2 + 0 =$ _____

3. $4 + 0 =$ _____

4. $5 + 0 =$ _____

5. $6 + 0 =$ _____

6. $8 + 0 =$ _____

7. $9 + 0 =$ _____

8. $7 + 0 =$ _____

9. $\begin{array}{r} 3 \\ +0 \\ \hline \end{array}$
10. $\begin{array}{r} 8 \\ +0 \\ \hline \end{array}$
11. $\begin{array}{r} 4 \\ +0 \\ \hline \end{array}$
12. $\begin{array}{r} 0 \\ +0 \\ \hline \end{array}$
13. $\begin{array}{r} 5 \\ +0 \\ \hline \end{array}$
14. $\begin{array}{r} 2 \\ +0 \\ \hline \end{array}$

Chapter 3 Addition

Compose Numbers

Many pairs of numbers added, have the same answer.

EXAMPLE: How many pairs of numbers added, equal 4?
$0 + 4 = 4$
$1 + 3 = 4$
$2 + 2 = 4$
$3 + 1 = 4$
$4 + 0 = 4$

EXAMPLE: How many pairs of numbers added, equal 5?
$0 + 5 = 5$
$1 + 4 = 5$
$2 + 3 = 5$
$3 + 2 = 5$
$4 + 1 = 5$
$5 + 0 = 5$

How many pairs of numbers added, equal 3?

1. Color 3 apples red and 0 apples blue.

3. Color 2 apples red and 1 apple yellow.

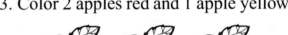

2. Color 1 apple red and 2 apples green.

4. Color 0 apples blue and 3 apples green.

Compose Numbers

How many pairs of numbers added, equal 7?

7 + 0 = 7	5 + 2 = 7	3 + 4 = 7	1 + 6 = 7
6 + 1 = 7	4 + 3 = 7	2 + 5 = 7	0 + 7 = 7

1. Color 7 frogs green and 0 frogs purple.

2. Color 6 frogs red and 1 frog brown.

3. Color 5 frogs blue and 2 frogs brown.

4. Color 4 frogs green and 3 frogs yellow.

5. Color 3 frogs red and 4 frogs purple.

Chapter 3   Addition

Counting On

Counting on is starting at a number then counting forward.

EXAMPLE: Start on 4. Count forward 2 more.
Say 4, 5, 6.
So, 4 + 2 = 6

I have two dogs. My dogs had two puppies.

I say 2, 3, 4.

So, 2 + 2 = 4.

Start with the first number. Count on. Write the answer.

1. 3 + 2 = _____
2. 1 + 3 = _____
3. 2 + 4 = _____
4. 6 + 1 = _____
5. 8 + 2 = _____

6. 3 + 5 = _____
7. 5 + 2 = _____
8. 4 + 3 = _____
9. 7 + 2 = _____
10. 2 + 6 = _____

11. 4 + 1 + 1 = _____
12. 2 + 3 + 1 = _____
13. 5 + 1 + 2 = _____
14. 3 + 5 + 1 = _____
15. 1 + 3 + 4 = _____

Add. Fill in the sentence with the answer.

1. I have 3 books. I get 1 more. I have _____ books.

2. I have 2 yellow birds. I get 3 more birds. I have _____ birds.

3. I have 4 peanuts. I get 5 more. I have _____ peanuts.

Start with the first number. Count on. Write the answer.

4.	3 +1	7.	8 +1	10.	4 +2	13.	7 +0	16.	6 +2	19.	2 +3
5.	3 +0	8.	7 +1	11.	4 +4	14.	9 +1	17.	5 +4	20.	3 +6
6.	2 +1 +1	9.	4 +1 +2	12.	6 +2 +1	15.	5 +0 +2	18.	3 +1 +2	21.	4 +3 +2

Doubles

Doubles are adding the same number two times.

EXAMPLE: Add the same number twice.
1 + 1 = 2
2 + 2 = 4

Add the doubles. Write your answer.

1. Color two bears brown. Color two bears black.

2. Color 3 bugs blue. Color 3 bugs green.

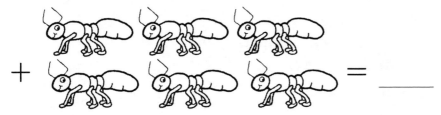

3. Color 4 bells orange. Color 4 bells purple.

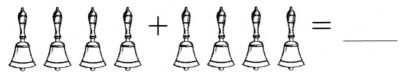

4. Color 5 roses yellow. Color 5 roses red.

Doubles

EXAMPLE: Doubles problem: $2 + 2 = 4$
Doubles plus one: $4 + 1 = 5$

EXAMPLE: Doubles problem: $3 + 3 = 6$
Doubles plus one: $6 + 1 = 7$

Add the doubles. Write your answer in both middle spaces.
Next, add one to the doubles answer. Write that answer.
The first one is done for you.

1. $1 + 1 = \underline{\ 2\ } \quad \underline{\ 2\ } + 1 = \underline{\ 3\ }$
2. $2 + 2 = \underline{\ \ \ \ } \quad \underline{\ \ \ \ } + 1 = \underline{\ \ \ \ }$
3. $3 + 3 = \underline{\ \ \ \ } \quad \underline{\ \ \ \ } + 1 = \underline{\ \ \ \ }$
4. $4 + 4 = \underline{\ \ \ \ } \quad \underline{\ \ \ \ } + 1 = \underline{\ \ \ \ }$
5. $5 + 5 = \underline{\ \ \ \ } \quad \underline{\ \ \ \ } + 1 = \underline{\ \ \ \ }$

Add the doubles. Write your answer.
Then add one more. Write your answer.
The first one is done for you.

6. $\begin{array}{r} 1 \\ +1 \\ \hline 2 \\ +1 \\ \hline 3 \end{array}$

7. $\begin{array}{r} 2 \\ +2 \\ \hline \\ +1 \\ \hline \end{array}$

8. $\begin{array}{r} 3 \\ +3 \\ \hline \\ +1 \\ \hline \end{array}$

9. $\begin{array}{r} 4 \\ +4 \\ \hline \\ +1 \\ \hline \end{array}$

10. $\begin{array}{r} 5 \\ +5 \\ \hline \\ +1 \\ \hline \end{array}$

Chapter 3   Addition

Making Tens

There were 8 birds in a tree. 2 more birds came to the tree. How many birds are there now?

8 + 2 = 10   10 = 1 ten

Tens	Ones
1	0

Add the numbers. Write the total.

1. 0 + 10 = _____
2. 1 + 9 = _____
3. 2 + 8 = _____
4. 3 + 7 = _____
5. 4 + 6 = _____
6. 5 + 5 = _____
7. 6 + 4 = _____
8. 7 + 3 = _____
9. 8 + 2 = _____
10. 9 + 1 = _____
11. 10 + 0 = _____

Making More Tens and Ones

Read the number. Color the number of blocks that is equal to the number.

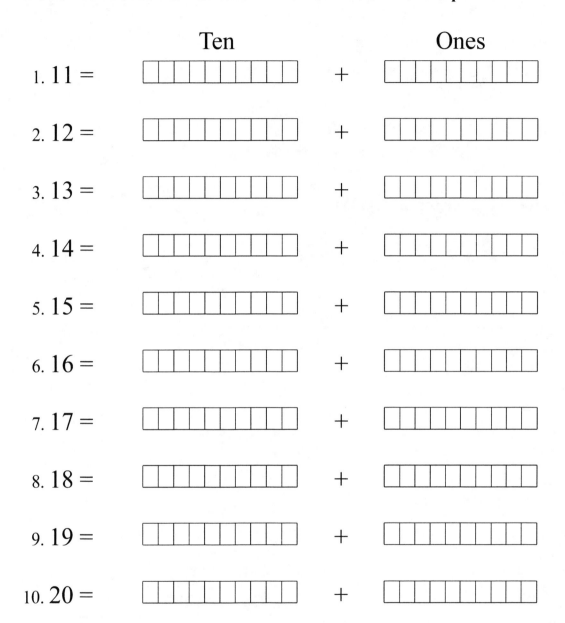

Read the problem. Add. Write how many tens and ones.

11. Alice has 6 beads. Her mom gave her 5 more. How many beads does Alice have in all?

Tens	Ones

12. Jim has 7 marbles. His dad gave him 8 more. How many marbles does Jim have in all?

Tens	Ones

Chapter 3 Addition

Adding in Any Order

Numbers can be added in any order. The sum will be the same.

EXAMPLE: 3 + 1 = 4 ♣ ♣ ♣ ★
 1 + 3 = 4 ♣ ★ ★ ★

EXAMPLE: 2 + 3 = 5 ♦ ♦ ♣ ♣ ♣
 3 + 2 = 5 ♦ ♦ ♦ ♣ ♣

Add the pairs of number sentences. Write the answers.

1. 6 + 2 = _____

   Color 6 blocks red.
   Color 2 blocks blue.

   2 + 6 = _____

   Color 2 blocks red.
   Color 6 blocks blue.

2. 3 + 4 = _____

   Color 3 blocks red.
   Color 4 blocks blue.

   4 + 3 = _____

   Color 4 blocks red.
   Color 3 blocks blue.

3. 5 + 3 = _____

   Color 5 blocks red.
   Color 3 blocks blue.

   3 + 5 = _____

   Color 3 blocks red.
   Color 5 blocks blue.

4. 1 + 7 = _____

   Color 1 block red.
   Color 7 blocks blue.

   7 + 1 = _____

   Color 7 blocks red.
   Color 1 block blue.

Adding Across and Down

Add across. Write the answer.

1. 12 + 1 = __13__
2. 14 + 3 = _____
3. 9 + 7 = _____
4. 6 + 8 = _____
5. 10 + 5 = _____

6. 7 + 5 = _____
7. 13 + 7 = _____
8. 11 + 3 = _____
9. 15 + 2 = _____
10. 12 + 7 = _____

Add down. Write the answer.

11. 4
    + 6
    ___
     10

12. 6
    + 2
    ___

13. 8
   + 10
    ___

14. 1
   + 17
    ___

15. 7
    + 3
    ___

16. 7
    + 6
    ___

17. 9
   + 11
    ___

18. 2
   + 13
    ___

19. 3
   + 12
    ___

20. 1
   + 19
    ___

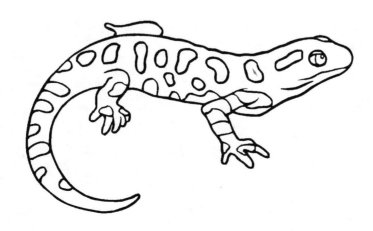

Chapter 3 Addition

Adding 3 Numbers

EXAMPLE: Solve: 3 + 2 + 4 = _____
First: Add 3 + 2.   3 + 2 = 5
Next: Add the 5 to the 4.   5 + 4 = 9
So, 3 + 2 + 4 = 9

EXAMPLE: Solve: 7 + 3 + 2 = _____
First: Add 7 + 3.   7 + 3 = 10
Next: Add the 10 to the 2.   10 + 2 = 12
So, 7 + 3 + 2 = 12

Add. Write the answer.

1. 2 + 5 + 1 = __8__
2. 11 + 1 + 2 = _____
3. 8 + 7 + 1 = _____
4. 4 + 3 + 3 = _____
5. 12 + 3 + 4 = _____

6. 9 + 1 + 1 = _____
7. 16 + 2 + 3 = _____
8. 10 + 7 + 2 = _____
9. 13 + 1 + 4 = _____
10. 6 + 2 + 7 = _____

11. 1
    + 6
    + 8

     15

12. 6
    + 2
    + 8

13. 4
    + 2
    + 7

14. 14
    + 2
    + 1

15. 11
    + 4
    + 4

72   Copyright © American Book Company

More Practice

Add. Write the answer.

1. $\begin{array}{r} 2 \\ +4 \\ +6 \\ \hline 12 \end{array}$

2. $\begin{array}{r} 6 \\ +3 \\ +1 \\ \hline \end{array}$

3. $\begin{array}{r} 3 \\ +3 \\ +4 \\ \hline \end{array}$

4. $\begin{array}{r} 7 \\ +7 \\ +1 \\ \hline \end{array}$

5. $\begin{array}{r} 4 \\ +3 \\ +7 \\ \hline \end{array}$

6. $\begin{array}{r} 4 \\ +9 \\ +3 \\ \hline \end{array}$

7. $\begin{array}{r} 3 \\ +9 \\ +5 \\ \hline \end{array}$

8. $\begin{array}{r} 6 \\ +6 \\ +2 \\ \hline \end{array}$

9. $\begin{array}{r} 1 \\ +2 \\ +9 \\ \hline \end{array}$

10. $\begin{array}{r} 5 \\ +5 \\ +5 \\ \hline \end{array}$

11. $\begin{array}{r} 2 \\ +1 \\ +9 \\ \hline \end{array}$

12. $\begin{array}{r} 1 \\ +7 \\ +9 \\ \hline \end{array}$

13. $\begin{array}{r} 4 \\ +12 \\ +2 \\ \hline \end{array}$

14. $\begin{array}{r} 7 \\ +11 \\ +2 \\ \hline \end{array}$

15. $\begin{array}{r} 4 \\ +18 \\ +2 \\ \hline \end{array}$

16. $\begin{array}{r} 2 \\ +17 \\ +1 \\ \hline \end{array}$

17. $\begin{array}{r} 17 \\ +1 \\ +1 \\ \hline \end{array}$

18. $\begin{array}{r} 16 \\ +2 \\ +3 \\ \hline \end{array}$

19. $\begin{array}{r} 12 \\ +3 \\ +3 \\ \hline \end{array}$

20. $\begin{array}{r} 18 \\ +1 \\ +1 \\ \hline \end{array}$

Chapter 3   Addition

Add. Write the answer.

21. 13 + 5 = _____

22. 12 + 6 = _____

23. 8 + 11 = _____

24. 3 + 14 = _____

25. 2 + 17 = _____

26. 4 + 12 = _____

27. 7 + 10 = _____

28. 5 + 15 = _____

Add. Match the numbers on the bird to the color chart. Color the bird.

1 = yellow
2 = green
3 = red
4 = blue
5 = orange
6 = purple
7 = blue

Word Problems

Read the problem. Find the sum.
Write the answer.

1. Amy found 3 kittens at the park. She has 2 kittens at home.
   How many kittens does she have in all? ____

2. Bob has 2 books. His mom gave him 2 more books for his birthday.
   His Aunt gave him 1 more book for his birthday.
   How many books does Bob have now? ____

3. At the pet store, there are 3 cats and 4 dogs.
   Then 2 more cats were brought in.
   How many cats and dogs are in the pet store now? ____

4. Kim found 6 shells. Her brother found 7 shells.
   Her sister found 3 shells. How many shells are there in all? ____

5. John found 4 bugs. His friend, Bill, found 8 bugs.
   How many bugs do they have in all? ____

6. Rita found 8 pretty rocks. Her friend, Lisa, found 7 pretty rocks.
   How many pretty rocks did they find in all? ____

7. Todd has 4 crayons. His brother, Jason, has 5 crayons.
   Their sister, Betty, has 6 crayons. How many crayons are there in all? ____

8. Wade has 3 toy cars. Jake has 5 toy cars.
   Their friend Paulo has 9 toy cars. How many toy cars are there in all? ____

Chapter 3   Addition

Chapter 3 Review

1. Skip count by twos. Write the numbers.

2 ____ ____ ____

2. Write in the + and the = in the ◯.

   4 ◯ 2 ◯ 6

Add. Write the sum.

3. 6 + 0 = _____

4. 5 + 13 = _____

5. 2 + 12 = _____

6. 8 + 7 = _____

Circle the number sentences that equal 12.

7. a) 5 + 8   b) 2 + 10   c) 9 + 2

8. a) 3 + 3 + 6   b) 1 + 11 + 2   c) 4 + 4 + 3

9. a) 7 + 5   b) 2 + 1 + 7   c) 9 + 3 + 1

Add. Fill in the sentence with the answer.

10. I have 2 balls. I got 3 more for my birthday.
    How many balls do I have now?   _____

11. I have 6 marbles. I got 4 more marbles.
    How many marbles do I have now?   _____

Add down. Write your answer.

12. 5   13. 3   14. 1   15.   8   16. 14   17. 11
    + 2   + 2   + 4   + 10   + 4   + 5
    + 1   + 8   + 6   + 2   + 1   + 2
    ___   ___   ___   ____   ___   ___

Add across. Write your answer.

18. 10 + 7 = _____   23. 8 + 7 = _____

19. 17 + 3 = _____   24. 12 + 4 = _____

20. 19 + 1 = _____   25. 11 + 7 = _____

21. 16 + 2 = _____   26. 10 + 2 = _____

22. 9 + 9 = _____   27. 14 + 4 = _____

Read the problem. Find the sum.
Write the answer.

28. Jan found 2 pennies on the ground. She has 16 pennies at home.
    How many pennies does Jan have now? _____

29. Don has 8 rocks at home. He found 3 more rocks he likes.
    How many rocks does Don have now?

30. Jake has 2 hats. He got 1 more for his birthday.
    How many hats does Jake have in all?

Chapter 3   Addition

Chapter 3 Test

Read the directions and circle the right answer.

1. Skip count by fives. Which list of numbers is correct?

   A 15, 21, 25

   B 15, 20, 25

   C 15, 20, 26

2. Which number sentence is correct?

   A $8 + 9 = 17$

   B $8 + 9 = 16$

   C $8 + 9 = 15$

3. Which number sentence is equal to 9?

   A $5 + 3$

   B $6 + 2$

   C $6 + 3$

4. I have 4 rocks. I found 3 more rocks. How many rocks do I have in all?

   A 7

   B 6

   C 5

5. Add down. Which is the correct answer?

   $$\begin{array}{r} 8 \\ +2 \\ +3 \\ \hline \end{array}$$

   A 12

   B 13

   C 14

6. 9 + 3 + 5 =

★★★★★★★★★ + ★★★ + ★★★★★

   A 17

   B 16

   C 15

7. Jeff found 3 pennies. He has 15 pennies at home. How many pennies does Jeff have now?

   A 15

   B 16

   C 18

8. 12 + 4 =

   A 17

   B 16

   C 15

9. Skip count by tens. Which number list is correct?

   A 10, 20, 30, 31

   B 30, 40, 50, 60

   C 50, 60, 61, 70

10. Which sentence is true?

   A 3 + 5 = 8 and 5 + 3 = 9.

   B 3 + 5 = 9 and 5 + 3 = 9.

   C 3 + 5 = 8 and 5 + 3 = 8.

Chapter 4
Subtraction

This chapter covers the following Georgia Performance Standards:

| M1N | Number and Operations | M1N3b, c, d, e, f, h |

Skip Counting Backwards

EXAMPLE: Skip count **backwards** by twos.

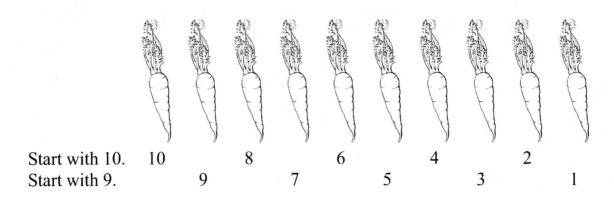

Start with 10.   10   8   6   4   2
Start with 9.   9   7   5   3   1

1. Below are 10 sandals. Skip count **backwards** by twos. Write the numbers.

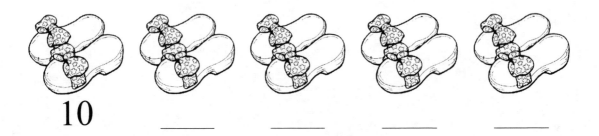

10   ___   ___   ___   ___

80   Copyright ©American Book Company

Skip Counting Backwards

EXAMPLE: Skip count **backwards** by fives.

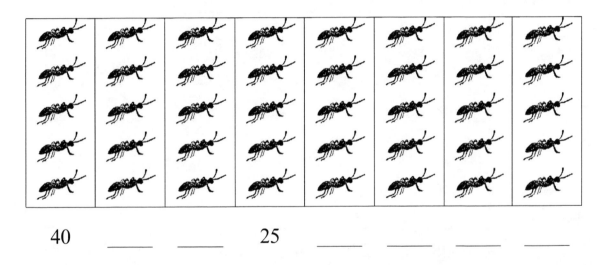

2. Skip count **backwards** by fives. Write the numbers.

40 _____ _____ 25 _____ _____ _____ _____

EXAMPLE: Skip count **backwards** by tens.

3. Skip count **backwards** by tens. Write the numbers.

60 _____ _____ _____ _____ _____

Chapter 4   Subtraction

Subtraction Sentences

The + sign means to add. $4 + 5 = 9$.

The − sign means to subtract. $9 - 4 = 5$.

Subtraction is the same as take away, minus, or how many less.

EXAMPLE:

Alex has 5 balloons.
Alex gives 2 balloons to his brother.
How many balloons does Alex have now?

5 balloons:

minus 2 balloons:

= 3 balloons

5 balloons − 2 balloons = 3 balloons $5 - 2 = 3$

Write in the − and the = in the ◯. The first one is done for you.

1. $6 - 2 = 4$

2. $8 \;\bigcirc\; 2 \;\bigcirc\; 6$

3. $9 \;\bigcirc\; 7 \;\bigcirc\; 2$

4. $4 \;\bigcirc\; 2 \;\bigcirc\; 2$

Subtraction Sentences

**Use the pictures. Write in the missing − and = signs.
Subtract the numbers. Write in the answer.**

1.    3

2.    ___

3.    ___

4.    ___

Subtract. Write the answer.

5. 7 − 3 = ___

6. 6 − 4 = ___

7. 3 − 2 = ___

8. 9 − 4 = ___

9. 8 − 6 = ___

Chapter 4 Subtraction

Subtracting All or Zero

When you subtract zero from any number, the answer will be the same as the first number in the sentence.

EXAMPLE: Subtract zero from 8.
$8 - 0 = 8$

EXAMPLE: Subtract zero from 2.
$2 - 0 = 2$

Subtract. Write the answer.

1. $12 - 0 = $ _____

2. $9 - 0 = $ _____

3. $7 - 0 = $ _____

4. $18 - 0 = $ _____

5. $10 - 0 = $ _____

6. $6 - 0 = $ _____

7. $1 - 0 = $ _____

8. $13 - 0 = $ _____

When you subtract a number from the same number, the answer will be zero.

EXAMPLE: Subtract 11 from 11.
$11 - 11 = 0$

EXAMPLE: Subtract 5 from 5.
$5 - 5 = 0$

Subtract. Write the answer.

9. $14 - 14 = $ _____

10. $7 - 7 = $ _____

11. $12 - 12 = $ _____

12. $3 - 3 = $ _____

13. $16 - 16 = $ _____

14. $4 - 4 = $ _____

15. $1 - 1 = $ _____

16. $19 - 19 = $ _____

Decompose Numbers Up to 10

You can make many subtraction sentences from a single number.

You can subtract all the numbers less than a number, and you can subtract the same number from itself.

EXAMPLE: Subtract all the numbers you can from 10.

$10 - 0 = 10$
$10 - 1 = 9$
$10 - 2 = 8$
$10 - 3 = 7$
$10 - 4 = 6$
$10 - 5 = 5$
$10 - 6 = 4$
$10 - 7 = 3$
$10 - 8 = 2$
$10 - 9 = 1$
$10 - 10 = 0$

There are 11 subtraction sentences you can make from the number 10.

Subtract the numbers from 8.

1. $8 - 0 =$ _____
2. $8 - 1 =$ _____
3. $8 - 2 =$ _____
4. $8 - 3 =$ _____
5. $8 - 4 =$ _____
6. $8 - 5 =$ _____
7. $8 - 6 =$ _____
8. $8 - 7 =$ _____
9. $8 - 8 =$ _____

EXAMPLE: Subtract by crossing out the number in the problem.

$6 - 2 = 4$

6

minus 2

= 4

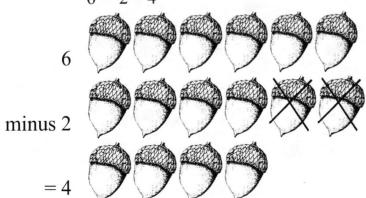

Chapter 4 Subtraction

Subtract. Cross out the number in the picture.

1. 5 – 2 = ___

2. 7 – 5 = ___

3. 4 – 3 = ___

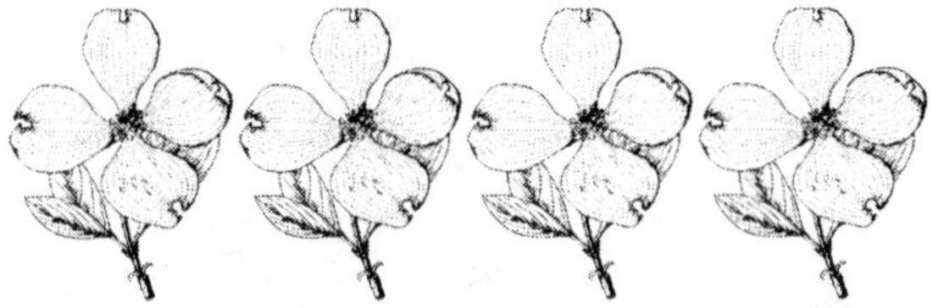

4. 6 – 1 = ___

5. 8 – 3 = ___

6. 3 – 1 = ___

7. 9 – 5 = ___

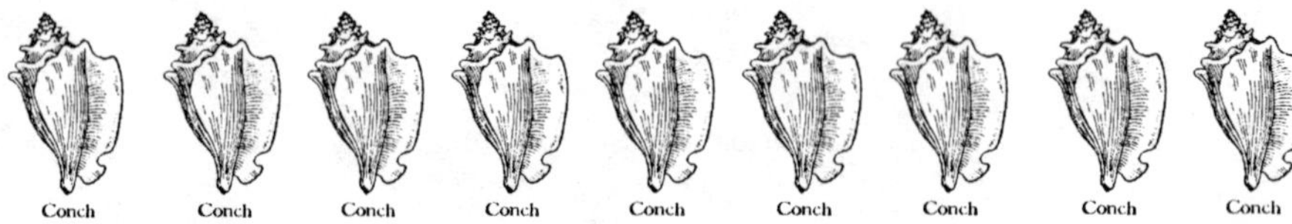

Counting Back

Counting back is starting at a number then counting backwards.

EXAMPLE: Start on 7. Count backward 2.
Say 7, 6, 5.
So, 7 − 2 = 5

I have 6 kittens. I give 4 kittens away.

I say 6, 5, 4, 3, 2.

So, 6 − 4 = 2.

Start with the first number. Count backwards. Write the answer.

1. 3 − 3 = _____
2. 11 − 5 = _____
3. 14 − 7 = _____
4. 9 − 2 = _____
5. 8 − 1 = _____

6. 5 − 3 = _____
7. 10 − 9 = _____
8. 17 − 4 = _____
9. 13 − 6 = _____
10. 4 − 3 = _____

11. 14 − 9 = _____
12. 7 − 6 = _____
13. 16 − 8 = _____
14. 4 − 2 = _____
15. 6 − 5 = _____

Chapter 4 Subtraction

Subtract. Fill in the sentence with the answer.

1. I have 7 buttons. I lose 1 of them. I now have _____ buttons.

2. I have 11 shells. I give 3 away. I now have _____ shells.

3. I see 3 birds. I see 2 birds fly away. I now see _____ bird(s).

Start with the first number. Count backwards. Write the answer.

4. 10   6. 11   8. 12   10. 13   12. 8   14. 19
   − 3   − 7   − 5   − 3   − 4   − 6

5.   4   7. 14   9. 16   11. 12   13. 15   15. 6
   − 0   − 14   − 4   − 1   − 9   − 2

Fact Families

Fact families are three numbers related to each other through addition and subtraction.

EXAMPLE: Are the numbers 1, 3, and 4 a fact family?

Test the numbers to see if they are a fact family.
1 + 3 = 4   3 + 1 = 4
4 − 1 = 3   4 − 3 = 1

Yes, all the number sentences are true. So 1, 3, and 4 is a fact family.

Look at each fact family and fill in the blanks under the fact family.

1. (2, 9, 7)

    2 + 7 = ____   9 − 2 = ____

    7 + ____ = 9   9 − ____ = 2

2. (7, 3, 4)

    3 + 4 = ____   7 − 3 = ____

    4 + ____ = 7   7 − ____ = 3

3. (3, 3, 6)

    3 + 3 = ____   6 − 3 = ____

    3 + ____ = 6   6 − ____ = 3

4. (1, 7, 6)

    1 + 6 = ____   7 − 1 = ____

    6 + ____ = 7   7 − ____ = 6

Chapter 4   Subtraction

Subtracting Across and Down

Subtract across. Write the answer.

1. 11 – 5 = __6__
2. 7 – 3 = _____
3. 12 – 4 = _____
4. 5 – 1 = _____
5. 8 – 2 = _____

6. 15 – 6 = _____
7. 13 – 8 = _____
8. 10 – 9 = _____
9. 17 – 17 = _____
10. 9 – 7 = _____

Subtract down. Write the answer.

11.	11 – 7 ――― 4	13.	14 – 6 ―――	15.	19 – 5 ―――	17.	7 – 2 ―――	19.	16 – 8 ―――	21.	4 – 2 ―――
12.	15 – 9 ―――	14.	3 – 1 ―――	16.	18 – 3 ―――	18.	12 – 8 ―――	20.	10 – 1 ―――	22.	11 – 1 ―――

Color the fruit any colors you want.

Comparing

EXAMPLE: Draw a set with 1 fewer. (Fewer means less.)

This set has 5. ♣ ♣ ♣ ♣ ♣

This set has 1 fewer, 4. ♣ ♣ ♣ ♣

$5 - 1 = 4$

Count the number in each set.
Draw another set with <u>1</u> fewer.

1.

2.

3.

4.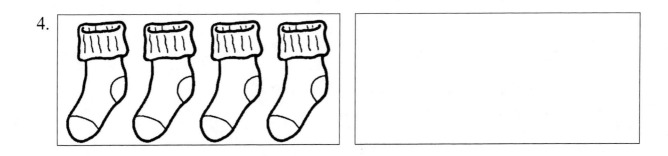

Chapter 4 Subtraction

Count the number in the set.
Draw another set with 2 fewer.

1.

Count the number in the set.
Draw another set with 3 fewer.

2.

Count the number in the set.
Draw another set with 4 fewer.

3.

Read the problem. Fill in the number sentence.

4. Jamal has 7 toy cars. His cat pushes 2 away. How many toy cars does Jamal have left?

   $7 - \underline{} = \underline{}$

5. Emma has 4 toy bears. She gives 1 to her sister. How many toy bears does Emma have left?

   $\underline{} - 1 = \underline{}$

6. Lee has 5 pencils. He gives 2 pencils to his friend. How many pencils does Lee have left?

   $\underline{} - \underline{} = \underline{}$

More Practice

Subtract across. Write the answer.

1. 17 − 2 = __15__
3. 12 − 6 = _____

2. 9 − 4 = _____
4. 8 − 7 = _____

Subtract down. Write the answer.

| 5. 15
− 7
――
8 | 6. 10
− 4
―― | 7. 7
− 1
―― | 8. 10
− 2
―― | 9. 13
− 8
―― | 10. 14
− 7
―― |

Subtract each problem. Match the answer to the color chart. Color the picture.

| 1 = red | 3 = blue | 5 = black | 7 = orange |
| 2 = brown | 4 = green | 6 = purple | 8 = yellow |

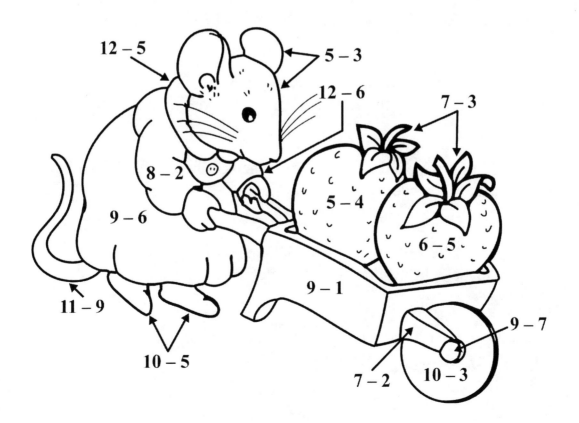

Chapter 4   Subtraction

Word Problems

Read the problem. Write the answer.

1. Rick has 5 balloons. He gives 2 balloons to his little brother.
   How many balloons does Rick have left?   _____

2. Kara found 6 pennies. She lost 2 pennies on the way home.
   How many pennies does Kara have left?   _____

3. The pet store has 18 orange fish. They sold 6 of the fish.
   How many orange fish are left?   _____

4. Abby has 9 hair ribbons. She gave 3 hair ribbons to her sister.
   How many hair ribbons does Abby have left?   _____

5. A robin has 4 eggs. The eggs hatch. When the baby birds are big,
   2 fly away. How many more birds still live at home?   _____

6. Kirk went fishing with his Uncle. Kirk got 3 fish.
   1 fish jumped back in the lake. How many fish did Kirk take home?   _____

7. Cora has 12 new pencils. She gives 3 pencils to her friend, Alena.
   How many pencils does Cora have left?   _____

8. Jake has 17 jelly beans. He gives 9 jelly beans to his dad.
   How many jelly beans does Jake have left?   _____

9. Sheldon has 20 toy cars. He gives 10 cars to his brother, Mark.
   How many cars does Sheldon have left?   _____

Chapter 4 Review

1. Skip count backwards by twos. Write the numbers.

Tiger lily   Tiger lily   Tiger lily   Tiger lily   Tiger lily   Tiger lily
12   _____   _____   _____   _____   _____

2. Write in the − and the = in the ◯.

   18 ◯ 4 ◯ 14

Subtract. Write the answer.

3. 12 − 2 = _____

4. 10 − 4 = _____

5. 8 − 7 = _____

6. 17 − 5 = _____

Circle the number sentences that equal 8.

7. a) 13 − 3   b) 12 − 4   c) 8 − 8

8. a) 8 − 0   b) 13 − 6   c) 9 − 8

9. a) 19 − 8   b) 16 − 8   c) 10 − 3

Read the problem. Write the answer.

10. I have 7 shells. I gave 3 away.
    How many shells do I have now?   _____

11. I have 10 rocks. I throw 5 in the lake.
    How many rocks do I have now?   _____

Chapter 4   Subtraction

Subtract down. Write the answer.

12. 14   13. 10   14. 7   15. 4   16. 18   17. 12
    − 4   − 6   − 2   − 1   − 9   − 9

Subtract across. Write the answer.

18. 13 − 8 = _____   23. 12 − 6 = _____

19. 9 − 8 = _____   24. 17 − 9 = _____

20. 16 − 4 = _____   25. 6 − 3 = _____

21. 8 − 7 = _____   26. 10 − 7 = _____

22. 6 − 4 = _____   27. 11 − 7 = _____

Read the problem. Write the answer.

28. Zach found 6 orange leaves, but 2 of the leaves blew away.
    How many leaves does Zach have left?   _____

29. Jan has 8 balloons. She gave 6 of them to friends.
    How many balloons does Jan have left?   _____

30. Hal has 7 toy cars. His cat pushed 4 of the cars out of his room.
    How many toy cars does Hal have left in his room?   _____

Chapter 4 Test

Read the directions and circle the right answer.

1. Skip count backwards by tens. Which list of numbers is correct?

    A 60, 55, 50

    B 60, 50, 40

    C 60, 50, 45

2. Which number sentence is correct?

    A $18 - 6 = 12$

    B $11 - 5 = 5$

    C $12 - 7 = 6$

3. Which number sentence is equal to 14?

    A $17 - 5$

    B $18 - 5$

    C $19 - 5$

4. I have 8 toy bears. I give 3 bears away. How many toy bears do I have left?

    A 5

    B 6

    C 7

5. Subtract down. Which is the correct answer?

    $$\begin{array}{r} 18 \\ -\ 8 \\ \hline \end{array}$$

    A 14

    B 12

    C 10

Chapter 4   Subtraction

6. 17 − 9 =

♣♣♣♣♣♣♣♣♣♣♣♣♣♣♣♣♣ − ♣♣♣♣♣♣♣♣♣ =

   A 8

   B 7

   C 6

7. Marcus found 9 pennies. He used 5 pennies for gum. How many pennies does Marcus have left?

   A 4

   B 6

   C 8

8. 11 − 7 =

   A 8

   B 6

   C 4

9. Skip count backwards by fives. Which number list is correct?

   A 20, 15, 10

   B 20, 10, 0

   C 20, 14, 10

10. Which sentence is true?

   A 10 − 4 = 7 and 10 − 7 = 4.

   B 12 − 8 = 4 and 12 − 4 = 8.

   C 14 − 6 = 7 and 14 − 7 = 6.

Chapter 5
Number Patterns

This chapter covers the following Georgia Performance Standards:

| M1N | Number and Operations | M1N3a |
| M1N | Number and Operations | M1N4a, b |

Patterns on a Hundred Chart

You can find **patterns** on a hundred chart.

1	2	3	4	5	6	7	8	9	10
11	12	13	14	15	16	17	18	19	20
21	22	23	24	25	26	27	28	29	30
31	32	33	34	35	36	37	38	39	40
41	42	43	44	45	46	47	48	49	50
51	52	53	54	55	56	57	58	59	60
61	62	63	64	65	66	67	68	69	70
71	72	73	74	75	76	77	78	79	80
81	82	83	84	85	86	87	88	89	90
91	92	93	94	95	96	97	98	99	100

EXAMPLE: Look at the column under the number 4.
The first number is 4.
Add 10 to 4.
Now, you have 14.
Add 10 to 14.
Now, you have 24.
Add 10 to 24.
Now, you have 34.

This pattern started with the number 4 and counted by tens.

Chapter 5   Number Patterns

1	2	3	4	5	6	7	8	9	10
11	12	13	14	15	16	17	18	19	20
21	22	23	24	25	26	27	28	29	30
31	32	33	34	35	36	37	38	39	40
41	42	43	44	45	46	47	48	49	50
51	52	53	54	55	56	57	58	59	60
61	62	63	64	65	66	67	68	69	70
71	72	73	74	75	76	77	78	79	80
81	82	83	84	85	86	87	88	89	90
91	92	93	94	95	96	97	98	99	100

Start on the given number. Count forward by tens.
Write the missing numbers.

1. 7, ____, ____, ____, ____, ____, ____, ____

2. 28, ____, ____, ____, ____, ____, ____, ____

3. 12, ____, ____, ____, ____, ____, ____, ____

4. 3, ____, ____, ____, ____, ____, ____, ____

5. 26, ____, ____, ____, ____, ____, ____, ____

6. 9, ____, ____, ____, ____, ____, ____, ____

7. 11, ____, ____, ____, ____, ____, ____, ____

8. 5, ____, ____, ____, ____, ____, ____, ____

9. 24, ____, ____, ____, ____, ____, ____, ____

10. 12, ____, ____, ____, ____, ____, ____, ____

Even and Odd Patterns

Even numbers end in 0, 2, 4, 6, and 8.
Odd numbers end in 1, 3, 5, 7, and 9.

Even Number:

Even numbers can be grouped into pairs.
There are 8 dinosaurs in the group. 8 can be grouped into 4 pairs.

Odd number:

After making pairs, odd numbers have 1 leftover.
There are 7 Triceratops in the group. 7 can be grouped into 3 pairs with 1 leftover.

Chapter 5   Number Patterns

**Count how many. Write how many on the line.
Circle even or odd.**

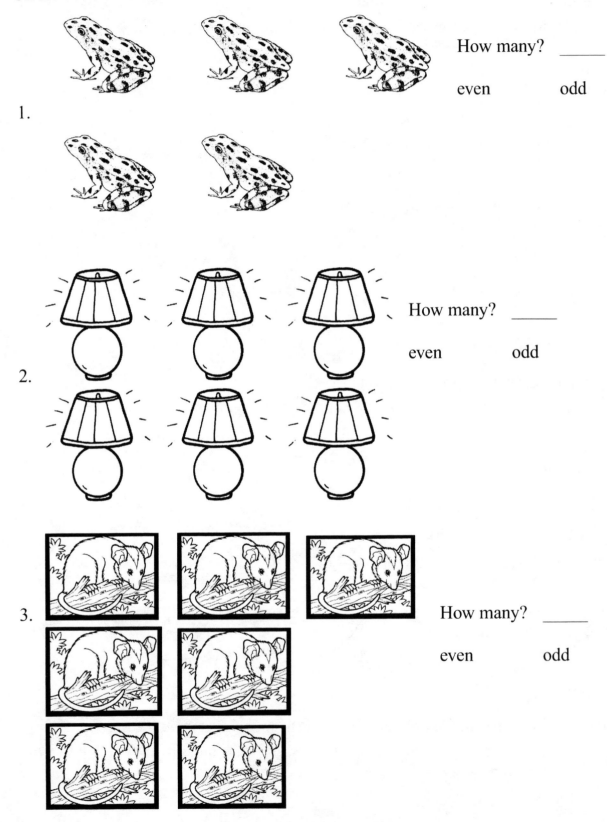

1. How many? _____

   even   odd

2. How many? _____

   even   odd

3. How many? _____

   even   odd

Even and Odd Patterns

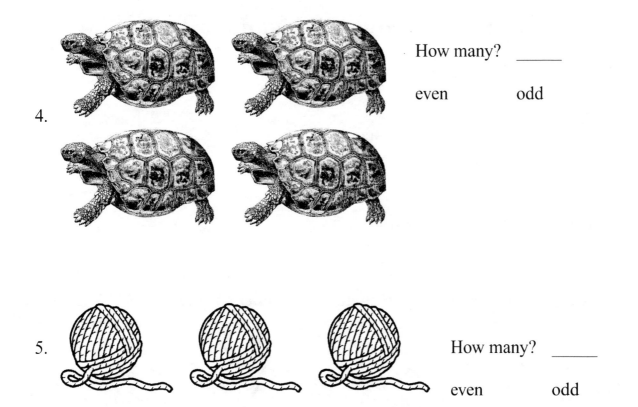

4. How many? _____

   even   odd

5. How many? _____

   even   odd

Color the even numbers yellow.
Color the odd numbers blue.

1	2	3	4	5	6	7	8	9	10
11	12	13	14	15	16	17	18	19	20
21	22	23	24	25	26	27	28	29	30
31	32	33	34	35	36	37	38	39	40

Chapter 5 Number Patterns

Using Skip Counting for Number Patterns

Skip counting is counting by 2's, 5's, or 10's. Skip counting can be done by any number.

EXAMPLE: Skip count by **2's**. Start at 2.
2, 4, 6, 8, 10.

EXAMPLE: Skip count by **2's**. Start at 1.
1, 3, 5, 7, 9, 11.

Skip count by 2's. Write the missing numbers under the berries.

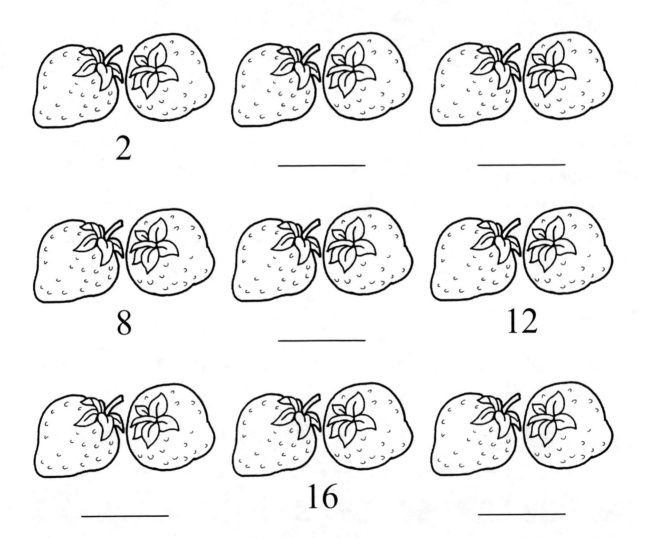

Using Skip Counting for Number Patterns

Skip count by 5's. Write the missing numbers under the hands.

5   _____   _____   20   _____

_____   35   _____   _____   50

55   _____   _____   70   _____

_____   85   _____   95   _____

Chapter 5   Number Patterns

Write the number that is one more than or one less than.

1. 17   18   _____
2. 12   13   _____
3. 24   25   _____
4. 87   88   _____
5. 62   63   _____
6. _____   9   10
7. _____   37   38
8. _____   78   79
9. _____   19   20
10. _____   42   43

Number Patterns

EXAMPLE: 22, ___, 42, 52

This pattern is counting by 10's.   22, 32, 42, 52

Use the number patterns you have learned to fill in the blanks. The patterns will count by 1's, by 2's, by 5's, or by 10's.

1. 14, 15, _____, 17

2. 10, _____, 30, 40

3. 22, 24, 26, _____

4. 30, 35, _____, 45

5. 44, _____, 48, 50

6. 55, _____, 57, 58

7. 60, 70, 80, _____

8. 5, 10, _____, 20

Chapter 5   Number Patterns

Use and Share Objects

EXAMPLE: John has 4 toy cars.
He wants to share the toy cars with his brother.
Each boy gets the same number of toy cars.
How many does each boy get?

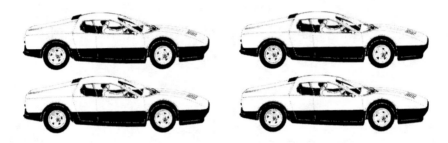

First: Put the toy cars into 2 equal groups.
Next: Count how many toy cars are in each group.
There are 2 toy cars in each group.
Each brother gets 2 toy cars to play with.

Read the problem. Write how many.
Draw the total number by group.

1. Ann has 6 balloons.
   Ann shares the balloons with 2 girl friends.
   How many balloons does each girl get?

   Answer: _____

Ann
Friend 1
Friend 2

Use and Share Objects

2. Karly has 8 pencils.
   Karly puts the same number of pencils in 2 boxes.
   How many pencils are in each box?

   Answer: _____

   | Box 1 |
   | Box 2 |

3. Jake's dad has 6 eggs.
   Jake, dad, and mom want eggs to eat.
   They each get equal number of eggs.
   How many will each person get?

   Answer: _____

   | Jake |
   | Jake's dad |
   | Jake's mom |

4. Cody has an ant farm with 15 ants in it.
   He wants to share his ants with 4 friends.
   Each of the 5 boys get an equal number of ants.
   How many ants will each boy get?

   Answer: _____

   | Al |
   | Friend 1 |
   | Friend 2 |
   | Friend 3 |
   | Friend 4 |

Chapter 5 Number Patterns

Chapter 5 Review

Start on the given number. Count forward by <u>tens</u>.

1. 8, _____, _____, _____, _____, _____,

Start on the given number. Count forward by <u>twos</u>.

2. 6, _____, _____, _____, _____, _____,

Start on the given number. Count forward by <u>fives</u>.

3. 5, _____, _____, _____, _____, _____,

Count how many. Write how many on the line. Circle even or odd.

4.

How many? _____

even   odd

Put an X over the <u>even</u> numbers.

5. | 1 | 2 | 3 | 4 | 5 | 6 | 7 | 8 | 9 | 10 |

Write the number that is <u>one</u> more than.

6. 15, 16, _____

7. 72, 73, _____

Chapter 5 Review

Write the number that is one less than.

8. _____, 20, 21

9. _____, 65, 66

Fill in the blanks of the number patterns below. The patterns may count by 1's, 2's, 5's, or 10's.

10. 15, 20, _____, 30

11. 4, 6, 8, _____

12. 16, 26, _____, 46

Read the problem. Write how many. Draw the total number by group.

13. Eva has 9 star stickers.
    Eva wants to share the stickers with two friends.
    How many will each person get?

    Answer: _____

Mary
Friend 1
Friend 2

Chapter 5 Test

Read the directions and circle the right answer.

1. Count forward by <u>tens</u>. Which number fills in the blank?

   23, 33, 43, _____

   A 53

   B 63

   C 73

2. Count forward by <u>fives</u>. Which number fills in the blank?

   30, 35, _____, 45

   A 50

   B 45

   C 40

3. Count how many. Is the number even or odd?

   A 3 books. An odd number.

   B 4 books. An even number.

   C 4 books. An odd number.

4. Write the number that is one more than.

   82, 83, _____

   A 84

   B 85

   C 86

5. Write the number that is one less than.

   _____, 47, 48

   A 45

   B 46

   C 47

6. The number pattern is counting by <u>twos</u>. Which number is missing?

   15, 17, _____, 21

   A 16

   B 18

   C 19

7. Ty's dog had 6 puppies. Ty and his sister name an equal number of puppies. How many puppies does each person name?

   A 3

   B 4

   C 6

Chapter 6
Fractions

This chapter covers the following Georgia Performance Standards:

| M1N | Number and Operations | M1N4c |

Equal Parts

You can cut an object into 2 **equal parts**.

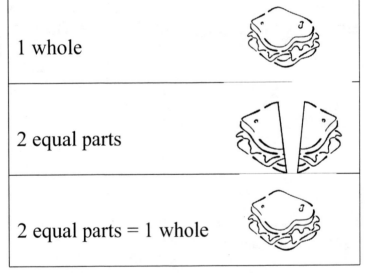

EXAMPLE:
| 1 whole |
| 2 equal parts |
| 2 equal parts = 1 whole |

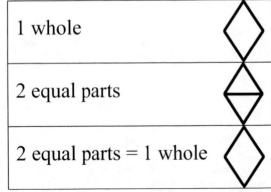

EXAMPLE:
| 1 whole |
| 2 equal parts |
| 2 equal parts = 1 whole |

Equal Parts

EXAMPLE:

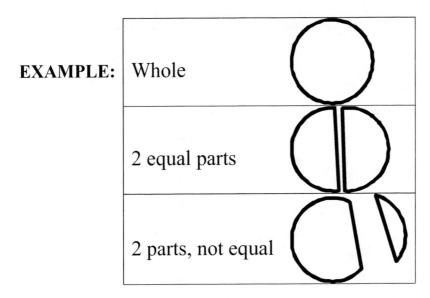

Look at each whole picture and the parts.
Are the parts equal or not equal? Circle the answer.

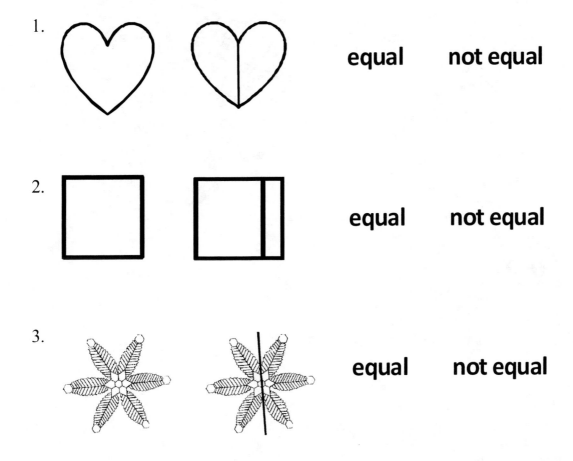

1. equal   not equal

2. equal   not equal

3. equal   not equal

Chapter 6   Fractions

Halves

Halves are 2 equal parts.

**Draw a line to divide each picture into 2 halves.
Color one half of each picture.**

1.

2.

3.

4.

5.

6.

Fourths

Fourths are 4 equal parts.

EXAMPLE:	1 whole	
	4 equal parts	
	4 equal parts = 1 whole	

Divide each picture into fourths. (4 equal parts)
Color <u>one</u> fourth.

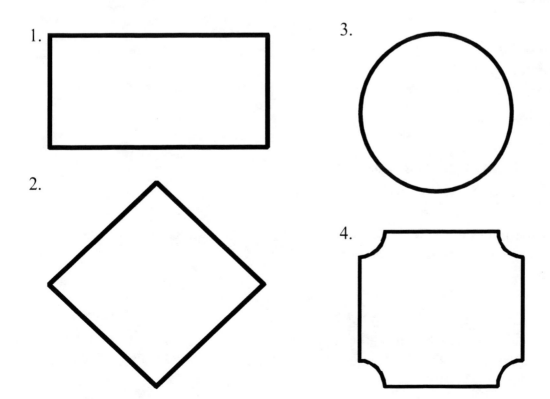

Chapter 6 Fractions

More Practice

Look at each drawing. Circle if it is showing halves or fourths.

1.

    halves   fourths

2.

    halves   fourths

3.

    halves   fourths

Read each problem. Circle the answer.

4. Jenny has 1 candy bar to share with her friend. How should Jenny cut the candy bar so each of them will get an equal share?   halves   fourths

5. Marco has one sandwich. He wants to share it with his 3 brothers. How should Marco cut the sandwich so each will get an equal share?   halves   fourths

Chapter 6 Review

Read each question. Circle the answer.

1. Elton and his 3 friends are sharing a pizza. How should Elton cut the pizza so each person gets an equal share?   halves   fourths

2. Lisa and her little sister are sharing an apple. Lisa divides the apple equally for the 2 of them. How many pieces will Lisa cut the apple?   halves   fourths

3. Mario has 8 marbles. He wants to give 2 to John, 2 to Jake, and 2 to Paul. How should Mario divide his marbles?   halves   fourths

Look at each picture and the parts.
Are the parts equal or not equal? Circle the answer.

4. 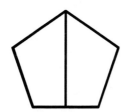   equal   not equal

Draw a line to divide the picture into 2 equal halves.
Color one half of the picture.

5.

Draw lines to divide the picture into 4 equal parts.
Color one fourth of the picture.

6.

Chapter 6 Fractions

Chapter 6 Test

Read the directions and circle the right answer.

1. How many parts are in the picture?

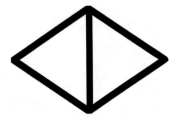

   A 1

   B 2

   C 4

2. A picture is divided into fourths. How many parts is the picture?

   A 1

   B 2

   C 4

3. Look at the picture. Which sentence best describes the picture?

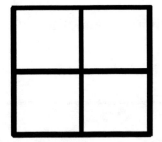

   A The picture is divided into equal halves.

   B The picture is divided into equal fourths.

   C The picture is divided into halves that are not equal.

Chapter 6 Test

4. Ira and his brother are sharing a cupcake.
   Ira cuts the cupcake so each gets one half.
   How many pieces did Ira cut the cupcake into?

   A 1

   B 2

   C 4

5. Kirby is sharing 8 cookies with John, Mason, and Rick.
   Each boy gets 2 cookies. How did Kirby divide the group of cookies?

   A one group

   B halves

   C fourths

6. Ned wants his mom to cut his sandwich into 4 equal parts.
   How should Ned's mom cut the sandwich?

   A Into 1 piece.

   B Into halves.

   C Into fourths.

Chapter 7
Money

This chapter covers the following Georgia Performance Standards:

| M1N | Number and Operations | M1N1e, f |

Pennies and Nickels

Penny

Nickel

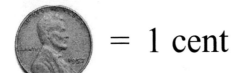

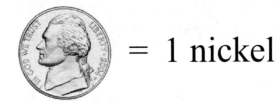

¢ = cent

Count the money. Write how much.

1.	(2 pennies)	=	_____ ¢
2.	(4 pennies)	=	_____ ¢
3.	(3 pennies)	=	_____ ¢
4.	(2 nickels)	=	_____ ¢
5.	(4 nickels)	=	_____ ¢
6.	(3 nickels)	=	_____ ¢
7.	(1 nickel, 1 penny)	=	_____ ¢
8.	(1 nickel, 2 pennies)	=	_____ ¢
9.	(2 nickels, 2 pennies)	=	_____ ¢
10.	(3 nickels, 3 pennies)	=	_____ ¢
11.	(4 nickels, 2 pennies)	=	_____ ¢
12.	(3 nickels, 4 pennies)	=	_____ ¢
13.	(2 nickels, 4 pennies)	=	_____ ¢
14.	(4 nickels, 1 penny)	=	_____ ¢

Chapter 7   Money

Dimes and Quarters

Dime

 = 1 dime

 = 10 cents

 = 10¢

Quarter

 = 1 quarter

 = 25 cents

 = 25¢

 = 1 dime = 10 cents = 10 ¢

 = 2 nickels = 10 pennies

 = 1 quarter = 25 cents = 25 ¢

 = 5 nickels = 25 pennies = 2 dimes + 1 nickel

Count the money. Write how much.

1.	2 dimes	=	_____ ¢
2.	4 dimes	=	_____ ¢
3.	6 dimes	=	_____ ¢
4.	2 quarters	=	_____ ¢
5.	3 quarters	=	_____ ¢
6.	2 quarters, 1 dime	=	_____ ¢
7.	1 quarter, 1 dime	=	_____ ¢
8.	1 quarter, 2 dimes	=	_____ ¢
9.	2 quarters, 2 dimes	=	_____ ¢
10.	3 quarters, 1 dime	=	_____ ¢
11.	1 quarter, 3 dimes	=	_____ ¢
12.	2 quarters, 3 dimes	=	_____ ¢

Chapter 7   Money

Counting Coins

To count money:
1st - Find the value of each coin.
2nd - Add.
3rd - Write the total.

EXAMPLE: Count the money.

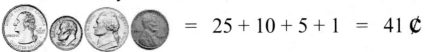 = 25 + 10 + 5 + 1 = 41 ¢

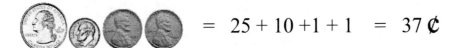 = 25 + 10 + 1 + 1 = 37 ¢

Write the name and the value under each coin.

1. Name _____   _____   _____   _____
   Value _____¢   _____¢   _____¢   _____¢

Find the value of each coin. Add.
Write the total.

2. = _____ = ___¢

3. = _____ = ___¢

4. = _____ = ___¢

Counting Coins

5. [quarter, dime, dime, dime, nickel] = _____ = ___¢

6. [dime, dime, dime, nickel, penny] = _____ = ___¢

7. [quarter, quarter, nickel, penny, penny] = _____ = ___¢

8. [quarter, quarter, quarter, penny] = _____ = ___¢

9. [quarter, dime, dime, nickel, penny] = _____ = ___¢

10. [quarter, quarter, dime, dime, dime] = _____ = ___¢

11. [quarter, nickel, nickel, nickel, penny] = _____ = ___¢

12. Look at the coins.

a) Put a line under the dime.
b) Put an x through the penny.
c) Circle the quarter.
d) Put a line above the nickel.

Fair Trades

In **fair trades**, we exchange coins for different coins, and we get back the same amount of money.

EXAMPLE: 1 quarter = 5 nickels 25 ¢ = 25 ¢
(fair trade) unfair trade

In **unfair trades**, we exchange coins for different coins, but we don't get back the same amount of money.

EXAMPLE: 1 quarter for 4 nickels 25 ¢ is **not** = to 20 ¢
fair trade (unfair trade)

Read each trade. Write the amounts of the trade.
Circle <u>fair</u> for a fair trade. Circle <u>unfair</u> for an unfair trade.

1.	1 quarter + 1 nickel for 3 dimes	_____ ¢ for _____ ¢	fair trade unfair trade
2.	2 quarters + 3 dimes for 6 dimes + 2 nickels	_____ ¢ for _____ ¢	fair trade unfair trade
3.	4 dimes + 10 pennies for 2 quarters	_____ ¢ for _____ ¢	fair trade unfair trade
4.	2 dimes + 2 nickels for 1 quarter and 1 penny	_____ ¢ for _____ ¢	fair trade unfair trade

Making Purchases

Dan and Sue are shopping. Dan has 82 cents. Sue has 57 cents. What can they buy?

Look at the toys and the price tags.
Can Dan or Sue buy the toy?
Circle yes or no for each toy.

		Dan has 82 cents.		Sue has 57 cents.	
1.	Airplane	Yes	No	Yes	No
2.	Balloon	Yes	No	Yes	No
3.	Dinosaur	Yes	No	Yes	No
4.	Car	Yes	No	Yes	No
5.	Bear	Yes	No	Yes	No
6.	Crayons	Yes	No	Yes	No

Balloon 25 cents

Toy Dinosaur 89 cents

Crayons 53 cents

Toy Bear 67 cents

Toy Car 97 cents

Toy Airplane 79 cents

Chapter 7   Money

Dollar Bills

 = 1 dollar = $1.00

 = 5 dollars = $5.00

 = 10 dollars = $10.00

 = 20 dollars = $20.00

Fun facts:

Dollars	=	Quarters	Dimes	Nickels	Pennies
1 dollar	=	4	10	20	100
5 dollars	=	20	50	100	500
10 dollars	=	40	100	200	1000
20 dollars	=	80	200	400	2000

Dollar Bills

Money that is one dollar or more is written with a $ sign before the number of dollars. The number of dollars is followed by a decimal and 2 zeros.

EXAMPLE: 5 dollars = $5.00
10 dollars = $10.00
15 dollars = $15.00

Count the money. Write how much.

1.  $5 + $1 = $_____.00

2.  $5 + $5 = $_____.00

3.  $5 + $1 + $1 = $_____.00

4.  $5 + $5 + $1 = $_____.00

5. $5 + $5 + $5 = $_____.00

6.  $10 + $20 = $_____.00

7.  $20 + $20 = $_____.00

8.  $10 + $10 = $_____.00

Chapter 7   Money

Fair Trades Using Bills

In **fair trades**, we exchange bills for different bills, and we get back the same amount of money.

EXAMPLE: one 20 dollar bill = four 5 dollar bills $20.00 = $20.00
                                    (fair trade)   unfair trade

In **unfair trades**, we exchange bills for different bills, but we don't get back the same amount of money.

EXAMPLE: one 20 dollar bill = three 5 dollar bills $20.00 = $15.00
                                    fair trade   (unfair trade)

Read each trade. Write the amounts of the trade.
Circle fair for a fair trade. Circle unfair for an unfair trade.

1.	One 10 dollar bill + 2 five dollar bills for one 20 dollar bill	$____.00 for $____.00 fair trade unfair trade
2.	One 20 dollar bill for one 10 dollar bill + five 1 dollar bills	$____.00 for $____.00 fair trade unfair trade
3.	Two 20 dollar bills for two 10 dollar bills + two 5 dollar bills and five 1 dollar bills	$____.00 for $____.00 fair trade unfair trade
4.	Three 5 dollar bills for one 10 dollar bill + five 1 dollar bills	$____.00 for $____.00 fair trade unfair trade

Making Purchases Using Bills

**Kim and Erin go to the pet store. Kim has $12.00. Erin has $15.00.
Look at the price of each pet. Can Kim and Erin buy the pet? Circle yes or no.**

	Pet	Price	Kim has $12.00	Erin has $15.00
1.	(otter)	$48.00	Yes No	Yes No
2.	(duck)	$12.00	Yes No	Yes No
3.	(rabbit)	$10.00	Yes No	Yes No
4.	(toucan)	$95.00	Yes No	Yes No
5.	(mouse)	$1.00	Yes No	Yes No
6.	(fish)	$5.00	Yes No	Yes No
7.	(parrot)	$27.00	Yes No	Yes No
8.	(puppy)	$15.00	Yes No	Yes No

Chapter 7 Review

Count the money. Write how much.

1.	quarter, nickel, nickel, penny	= _____	=	____¢
2.	dime, dime, dime, nickel	= _____	=	____¢
3.	quarter, quarter, penny, penny	= _____	=	____¢
4.	$5 bill, $1 bill	= _____	=	$___.00
5.	$10 bill, $5 bill	= _____	=	$___.00
6.	$20 bill, $1 bill	= _____	=	$___.00

Read each trade. Write the amounts of the trade.
Circle fair for a fair trade. Circle unfair for an unfair trade.

7.	2 quarters + 2 dimes for 7 dimes	____¢ fair trade	for	____¢ unfair trade
8.	One 10 dollar bill for one 5 dollar bill + five 1 dollar bills	____¢ fair trade	for	____¢ unfair trade
9.	3 dimes + 10 pennies for 1 quarter + 3 dimes	____¢ fair trade	for	____¢ unfair trade
10.	One 20 dollar bill for one 10 dollar bill + one 5 dollar bill	____¢ fair trade	for	____¢ unfair trade

11. Look at the coins.

a) Put a line over the penny.
b) Circle the dime.
c) Put an X through the quarter.
d) Put a line under the nickel.

12. John goes to the grocery store. He has 70 cents. Circle the items John can buy.

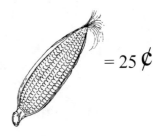

Fill in the chart.

13.	One 1 dollar bill =	_____ quarters
14.	One quarter =	_____ nickels
15.	One 20 dollar bill =	_____ 10 dollar bills
16.	Two dimes =	_____ pennies

Chapter 7   Money

Chapter 7 Test

Read the directions and circle the right answer.

1. Count the money. How much is it?

   A   31¢

   B   56¢

   C   76¢

2. Count the money. How much is it?

   A   40¢

   B   45¢

   C   50¢

3. Count the money. How much is it?

   A $8.00

   B $6.00

   C $4.00

4. Count the money. How much is it?

   A $25.00

   B $30.00

   C $40.00

5. Count the money. How much is it?

A $4.00

B $17.00

C $25.00

6. Alex goes to the toy store. He has $15.00. What can he buy?

7. Emma goes to the music store. She has $50.00 to spend. What can she buy?

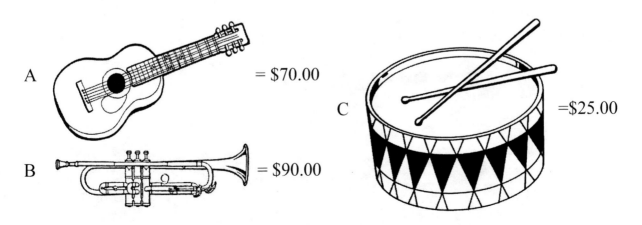

Chapter 8
Measurement

This chapter covers the following Georgia Performance Standards:

| M1N | Number and Operations | M1M1a, b, c |

Compare Length

EXAMPLE:	What are other words for shorter and longer?
Shorter:	smaller
Longer:	bigger, taller

EXAMPLE: Which is taller?

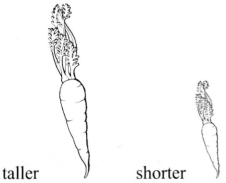

taller   shorter

EXAMPLE: Which is the shorter piece of rope?

shorter

longer

138   Copyright ©American Book Company

Compare Length

Draw a line under the shortest objects.

1.
2.
3.
4.
5.

6. Number the blocks from shortest to longest: 1, 2, and 3.
   1 is the shortest. 3 is the longest.

Chapter 8   Measurement

Measure Length Using Nonstandard Units

Use a small paperclip to measure length.

EXAMPLE: Use a small paper clip to measure the long side of the box.

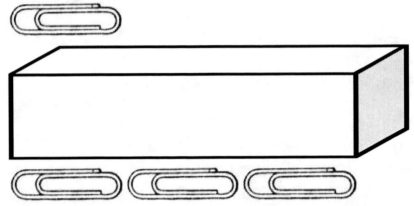

The long side of the box is about 3 paper clips long.

Measure the picture in each row using a small paper clip. Write how many. (If you do not have a small paper clip, use a small scrap of paper. Cut or tear the paper to be the same length as the small paper clip.)

1. About _____ paper clips tall.
2. About _____ paper clips wide.
3. About _____ paper clips tall.
4. About _____ paper clips wide.

Measure Length with a Tool

You can make your own measuring tool using a piece of paper and the wide part of your thumb.

Make a mark one thumb width from the edge of a piece of paper. Then make 19 more marks.

You now have a piece of paper with 20 thumb width marks.

(Do not draw the thumb, just make a mark for measuring.

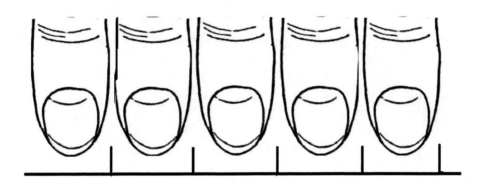

Use the ruler you just made to measure the list of things below.

1. How wide is your hand?   My hand is about _____ thumb widths wide.
2. How wide is your shoe?   My shoe is about _____ thumb widths wide.
3. How tall is your glue bottle?   My glue bottle is about _____ thumb widths.
4. How long is your red crayon?   My red crayon is about _____ thumb widths.
5. How long is your pencil?   My pencil is about _____ thumb widths.

Measure the boxes below with your thumb ruler.

6. ▭   7. ▭   8. ▭   9. ▭

6. The box is about _____ thumb widths long.
7. The box is about _____ thumb widths long.
8. The box is about _____ thumb widths long.
9. The box is about _____ thumb widths long.

Chapter 8   Measurement

Inches and Centimeters

Use a ruler to measure the small paper clip.

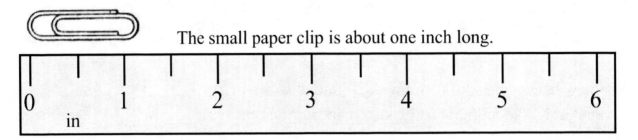

Below is 1 inch and 1 centimeter. Centimeters are another way to measure things.

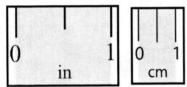

There are 12 inches in a foot. A ruler is 1 foot.
There are about 30 centimeters in a foot.

Measure the pictures below. Use inches and centimeters.

1. How long is the frog?

    inches _____   centimeters _____

3. How wide is the saucer?

    inches _____   centimeters _____

2. How long is the bug?

    inches _____   centimeters _____

4. How long is the leaf, stem to tip?

    inches _____   centimeters _____

Compare Weight

EXAMPLE: Which weighs **more**?

 or

The horse, of course!

<u>Circle</u> the one that weighs <u>more</u>.
<u>Color</u> the one that weighs <u>less</u>.

Chapter 8   Measurement

Below is a picture of a **balance scale**. It is used to measure the weight of one thing compared to another.

If the balance tips, the object that is **higher is lighter**.

The object that is **lower is heavier**.

If the balance **doesn't tip, they weigh the same**.

You do not need a balance to figure out which things weigh more.

You can be the balance! Put two objects, one in each hand.

Can you feel the difference in weight? Which one weighs more?

Which one weighs less?

Try it using these pairs of objects.

a) a pencil and a bottle of glue

b) a book and a box of crayons

c) a piece of paper and one crayon

d) a shoe and a tablet

Try this at home. Use any two objects and see if you can tell which one weighs more.

Measure Weight Using Nonstandard Units

You can measure weight by using one object to compare to all others.

EXAMPLE: Use a box of crayons to compare weight to other objects.
Compare a book to a box of crayons. The book weighs more.
Compare a pencil to a box of crayons. The crayons weigh more.
The order of the 3 objects by weight is: pencil, box of crayons, book.

Measure Weight Using Nonstandard Units

Use a box of crayons to compare weights.
If you don't have the object, make your best guess.
Put a 1 under the lightest object.
Put a 2 under the middle weight object.
Put a 3 under the heaviest object.

Chapter 8   Measurement

Compare Capacity

EXAMPLE: Which holds more? Which holds less?

The basket with 6 kittens holds more. The basket with 3 birds holds less.

Circle the one that holds **more**. **Underline** the one that holds **less**.

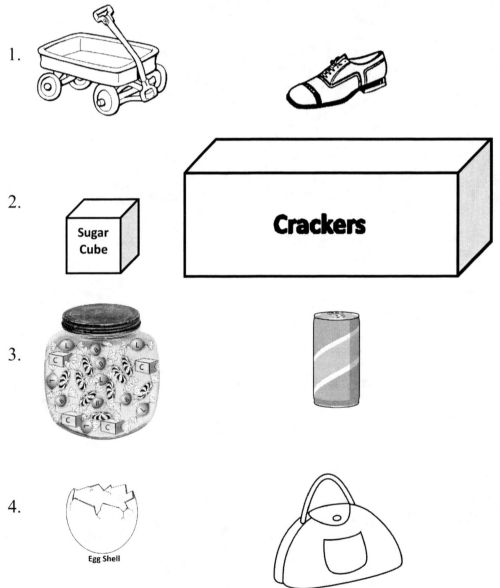

1.
2.
3.
4.

Measure Capacity Using Nonstandard Units

EXAMPLE: How many glasses of water will fit into the pail?

1. How many cupcakes will fit in the box?

 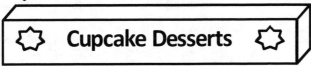

_____ cupcakes will fit in the box.

2. How many eggs will fit in the carton?

 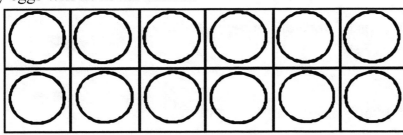

_____ eggs fit into the carton.

3. How many hands will fit into the group of mittens?

_____ hands fit into the group of mittens.

Measure Capacity with a Tool

Fun Facts:

1 cup

2 cups = 1 pint

2 pints = 1 quart

4 quarts = 1 gallon

Use the chart above to answer the questions. Write your answer.

1. How many cups of juice are there in 1 pint?

   There are _____ cups of milk in 1 pint.

2. How many quarts of milk are in 1 gallon?

   There are _____ quarts of milk in 1 gallon.

3. How many pints of water are in 1 quart?

   There are _____ pints of water in 1 quart.

4. How many cups of tea are in 1 pint?

   There are _____ cups of tea in 1 pint.

Chapter 8 Review

Draw a line under the longer ropes. Circle the shorter ropes.

1.
2.
3.

Use a small paper clip and measure the objects. Write how many.

4. The bee is about _____ paper clips wide.

5. The spoon is about _____ paper clips long.

Use a ruler and measure the objects. Some are measured in inches, some in centimeters.

6. The nickel is about _____ inch(es) wide.

Chapter 8   Measurement

7.

The butterfly is _____ centimeters at its widest part.

8. Circle the object that weighs more.

9. Circle the object that holds more.

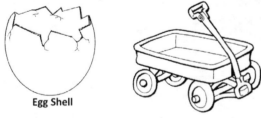
Egg Shell

10. How many quarts are in a gallon of milk?

There are _____ quarts in a gallon of milk.

Chapter 8 Test

Read the directions and circle the right answer.

1. Which is the <u>shortest</u> piece of rope?

2. Use a small paper clip to measure. Which object is about one paper clip wide?

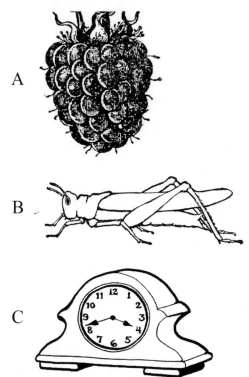

Chapter 8 Measurement

3. Which object weighs the most?

A

B

C

4. Which object holds the most?

A

B

C

Chapter 8 Test

5. How many cups of juice in a pint?

   A 1

   B 2

   C 4

6. How many pints of water in a quart?

   A 1

   B 2

   C 4

7. How many quarts of milk in a gallon?

   A 1

   B 2

   C 4

8. Which holds more?

   A One cup of milk.

   B One gallon of water.

   C Two quarts of juice.

9. Which holds more?

   A A truck full of wood.

   B A basket full of cookies.

   C A cup of tea.

1 cup

2 cups = 1 pint

2 pints = 1 quart

4 quarts = 1 gallon

Chapter 9
Time

This chapter covers the following Georgia Performance Standards:

| M1M | Measurement | M1M2a, b, c |

Tell Time to the Nearest Hour

A full day has 24 hours. A clock shows 12 hours, a half day.

The short hand on the clock shows the hour.

The long hand on the clock shows the minutes of the hour.

Each number is equal to 5 minutes. Count by fives for all 12 numbers on the clock.

1	2	3	4	5	6	7	8	9	10	11	12
5	10	15	20	25	30	35	40	45	50	55	60

There are 60 minutes in an hour.

The clock above shows 2 o'clock.
The hour hand is on the 2.
The minute hand is on the 12.

The clock above is a digital clock.
It shows 12 o'clock.
a.m. is for hours from midnight to noon.
p.m. is for hours from noon to midnight.

Tell Time to the Nearest Hour

EXAMPLE: Tell time to the nearest hour.

The minute hand is near the 12. The hour hand is on the 1.
Since the hour hand is on the 1, it is 1 o'clock.
1 o'clock can also be written as **1:00**.

**What time does the clock say? Round the time to the nearest hour.
Write the time as both ____ o'clock and __:00.**

1.  _____ o'clock   _____:00

2. _____ o'clock   _____:00

3.  _____ o'clock   _____:00

Now, you write the   o'clock   and   :00

4.    _____   _____

5.   _____   _____

Chapter 9   Time

Tell Time to the Nearest Half Hour

EXAMPLE: Rita's bedtime is **half past 7**. Her bedtime is **7:30**.
The hour hand is half past the 7. The minute hand is on the 6.
The 6 is equal to 30 minutes.
An hour has 60 minutes, so half past is at 30 minutes.
(Half of 60 is 30.)

**What time does the clock say? Round the time to the <u>nearest</u> half hour.
Write both the half past and the ____:30.**

1. The time is half past ____ .   ____:30

2. The time is half past ____ .   ____:30

3. The time is half past ____ .   ____:30

Now, you write   The time is half past ____ .   and   ____:30

4. _____   _____

5. _____   _____

Time for More Practice

**Read where the hour and minute hands of the clock are.
What time is it? Round to the nearest hour.**

1.  _____ o'clock.

4.  _____ o'clock.

2.  _____ o'clock.

5.  _____ o'clock.

3.  _____ o'clock.

6.  _____ o'clock.

**Read where the hour and minute hands of the clock are.
What time does the clock say? Round to the nearest half hour.**

7. half past _____

10.  half past _____

8. half past _____

11.  half past _____

9. half past _____

12.  half past _____

Chapter 9   Time

Sequence of Events

EXAMPLE:
> Jacob gets up at 6:00 a.m. to get ready for school.
> Jacob eats breakfast at 6:30 a.m.
> He gets on the bus at 7:00 a.m.
> He gets to school at 7:30 a.m.
> Jacob leaves school at 3:00 p.m.

Questions: What is Jacob doing at 6:30 a.m.?   He is eating breakfast.
              Where is Jacob at 2:00 p.m.?   He is at school.

Read the events of the day. Answer the questions.

> Dana walks to school at 7:00 a.m.
> She eats her lunch at 12:00 p.m.
> Dana leaves school at 3:00 p.m.

1. Where is Dana at 8:00 a.m.?   Dana is at _____ at 8:00 a.m.

2. What does Dana do at 12:00 p.m.?   Dana _____ at 12:00 p.m.

3. What time does Dana leave school?   Dana leaves school at _____ .

> Kaila gets up at 6:30 a.m. on Saturday.
> Kaila makes her bed at 7:00 a.m.
> Kaila eats her breakfast at 7:30 a.m.

4. What time does Kaila get up on Saturday?   Kaila gets up at _____ .

5. What does Kaila do at 7:30 a.m.?   Kaila _____ at 7:30 a.m.

6. What time does Kaila make her bed?   Kaila makes her bed at _____ .

Duration of Events

EXAMPLE: Which takes a **longer** time? Which takes a **shorter** time?
Doing the dishes.
Sleeping for one night.
Doing the dishes takes about 30 minutes.   **shorter**
Sleeping for one night takes 8 – 10 hours.   **longer**

EXAMPLE: Which takes a **longer** time? Which takes a **shorter** time?
Eating breakfast.
Brushing your teeth.
Eating breakfast takes 15 – 30 minutes.   **longer**
Brushing your teeth takes about 2 minutes.   **shorter**

Which takes a longer time? Which takes a shorter time?
Write <u>longer</u> by the one that takes more time.
Write <u>shorter</u> by the one that takes less time.

		Longer or shorter?
1.	Putting on your shoes.	_____
	Reading 5 pages.	_____
2.	Eating lunch.	_____
	Walking for 3 minutes.	_____
3.	Doing 12 jumping jacks.	_____
	Sleeping 9 hours.	_____
4.	Writing a thank you note.	_____
	Spending one day at school.	_____
5.	Walking for 5 minutes.	_____
	Singing the alphabet song.	_____
6.	Clapping your hands fast 10 times.	_____
	Writing a thank you note.	_____

Chapter 9   Time

Calendar

Fun Facts

There are 7 days in a week.
- 1   Sunday
- 2   Monday
- 3   Tuesday
- 4   Wednesday
- 5   Thursday
- 6   Friday
- 7   Saturday

There are 12 months in a year.
- 1   January
- 2   February
- 3   March
- 4   April
- 5   May
- 6   June
- 7   July
- 8   August
- 9   September
- 10   October
- 11   November
- 12   December

Read the question. Write the answer.

1. Write the month and date of your birthday. _____
2. What day does the school week start on? _____
3. Write the month and date of today. _____
4. What is the day after Friday? _____
5. What is the 7th month of the year? _____
6. What is the 1st month of the year? _____
7. What month comes after September? _____
8. How many days in one week? _____
9. How many months in a year? _____

Below is the January page from a calendar in the year 2010.

January, 2010

Sunday	Monday	Tuesday	Wednesday	Thursday	Friday	Saturday
					1 New Years Day	2
3	4	5	6	7	8	9
10	11	12	13	14	15 Band Concert	16
17	18	19	20	21	22	23
24 31	25	26 School Bake Sale	27	28	29	30

Use the calendar page above. Read the question. Write the answer.

1. What date is New Years Day? _____
2. What date is the School Bake Sale? _____
3. Which dates are all the Mondays? _____
4. What day of the week does the 20th fall on? _____
5. What day of the week is the school Band Concert? _____

Chapter 9 Review

What time does the clock say? Round the time to the nearest hour. Write the time as both ____ o'clock and __:00.

1.  _____ o'clock _____:00

2.  _____ o'clock _____:00

What time does the clock say? Round the time to the nearest half hour. Write both the half past and the __:30.

3.  The time is half past _____ . _____:30

4.  The time is half past _____ . _____:30

What time does the clock say? Round to the nearest hour.

5.  _____ o'clock.

6.  _____ o'clock.

Chapter 9 Review

What time does the clock say? Round to the nearest half hour.

7.  half past _____

8.  half past _____

Read the events of the day. Answer the questions.

Mario is shopping at the toy store at 9:00 a.m.
He goes to the park at 9:30 a.m.
Mario goes home at 12:00 p.m. (noon)

9. Where is Mario at 9:00 a.m.? Mario is at the _____ at 9:00 a.m.

10. What does Mario do at 12:00 p.m.? Mario _____ at 12:00 p.m.

11. Where does Mario go at 9:30 a.m.? Mario goes to the _____ at 9:30 a.m.

Chapter 9   Time

Which takes a longer time? Which takes a shorter time?
Write longer by the one that takes more time.
Write shorter by the one that takes less time.

                                                    Longer or shorter?

#		
12.	Putting on your shoes.	_____
	Eating your dinner.	_____
13.	Walking 5 minutes.	_____
	Sleeping one full night.	_____
14.	Reading 5 pages.	_____
	Singing the alphabet song.	_____

Read the question. Write the answer.

15. What is the day after Thursday? _____

16. What month comes after March? _____

17. Write the month and date of today. _____

18. What day comes after Monday? _____

19. What is the last month of the year? _____

Put the days of the week in the correct order. The first day is done for you.

    Mixed up days of the week   **Correct order of the days of the week**

20. Wednesday   # Sunday

21. Friday   _____

22. Sunday   _____

23. Monday   _____

24. Thursday   _____

25. Saturday   _____

26. Tuesday   _____

Chapter 9 Test

Read the directions and circle the right answer.

1. What time does the clock say? Round to the <u>nearest</u> hour.

   A   11 o'clock

   B   8 o'clock

   C   12 o'clock

2. What time does the clock say? Round to the <u>nearest</u> half hour.

   A   1:30

   B   5:30

   C   2:30

3. What day comes after Wednesday?

   A   Monday

   B   Tuesday

   C   Thursday

Chapter 9   Time

4. How many months are there in one year?

   A 12

   B 10

   C 8

5. Read the events of the day.

   > Dan gets up at 7:00 a.m.
   > He is in school by 8:00 a.m.
   > Dan goes to lunch at 12:30 p.m.

   Where is Dan at 10:00 a.m.?

   A Dan is still in bed.

   B Dan is on his way home.

   C Dan is at school.

6. Which takes the longest time?

   A getting a full night's sleep

   B brushing your teeth

   C eating your lunch

7. Which takes the shortest time?

   A putting on your shoes

   B taking a bath

   C walking 10 minutes

Chapter 9 Test

Look at the calendar below and answer the questions.

March, 2010

Sunday	Monday	Tuesday	Wednesday	Thursday	Friday	Saturday
	1	2	3	4 ★	5	6
7	8	9	10	11	12	13
14	15	16 ☺	17	18	19	20
21	22	23	24	25	26	27 Amy's Birthday
28	29	30	31			

8. What date is Amy's birthday?

   A March 16

   B April 27

   C March 27

9. What date has the star on it?

   A March 4

   B March 8

   C March 11

10. What day of the week is the last day of the month?

    A Monday

    B Tuesday

    C Wednesday

Chapter 10
Geometry

This chapter covers the following Georgia Performance Standards:

M1G	Geometry	M1G1a, b, c
		M1G2
		M1G3

Plane Figures

      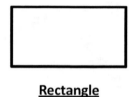   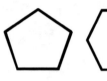

Triangle　　**Square**　　**Rectangle**　　**Pentagon**　　**Hexagon**

Above are pictures of shapes called **plane figures**.

The **sides** of a plane figure are lines.

The **corners** of a plane figure are pointy.

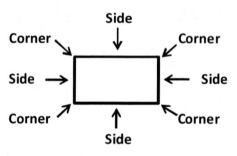

Plane Figure	Kinds	Number of Sides	Number of Corners
Triangle	△ ◸ △	3	3
Square	□	4	4
Rectangle	▭ ▯	4	4
Pentagon	⬠	5	5
Hexagon	⬡	6	6

168　　Copyright © American Book Company

Plane Figures

Fill in the chart below. Use a ruler or a straight edge to help you draw the figure.
Color the 3 sided shapes red.
Color the 4 sided shapes yellow.
Color the 5 sided shapes green.
Color the 6 sided shapes orange.

Name	Plane Figure	You draw the figure	Number of sides	Number of corners
Triangle	 △		3	
Triangle	 ◺			
Triangle	 △			3
Square	 □			
Rectangle	▭		4	
Rectangle	▯			
Pentagon	 ⬠			
Hexagon	 ⬡			6

Copyright © American Book Company

Chapter 10 Geometry

Solid Figures

The flat sides of a solid figure are called **faces**.

The points on a solid figure are called **corners**.

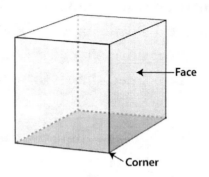

Cylinder — 2 faces, no corners
Cone — 1 face, 1 corner
Cube — 6 faces all the same, 8 corners
Rectangular Prism — 6 faces of three sizes, 8 corners

**Read the question. Choose the best solid figure.
Write the name of the solid figure.**

1. Which solid figure is like a drum?   _____

2. Which solid figure is like a book?   _____

3. Which solid figure is like this box?   _____

4. Which solid figure is like a trophy?  _____

5. Which solid figure is like a cake?  _____

Solid Figures

**Read the object. Draw a picture of the object.
Write the name of the solid figure it is like.**

Object	Draw the object	Name the solid figure
1. Can of soda.		_____
2. Box of cereal.		_____
3. Toy block.		_____
4. Ice cream cone.		_____

Look at the solid figure. Count how many faces and corners. Write the number.

	Solid Figure	Faces	Corners
5.	cylinder	_____	_____
6.	cone	_____	_____
7.	cube	_____	_____
8.	rectangular prism	_____	_____

Chapter 10   Geometry

Naming Faces of Solid Figures

A **cylinder** has **2 faces**. The top and bottom of a cylinder are faces.
The faces are shaped like circles.

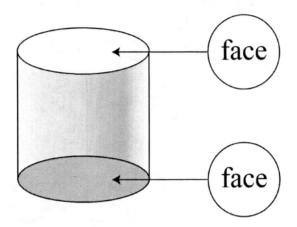

A **cone** has **1 face**. The bottom of a cone is the face.
A cone's face looks like a circle.

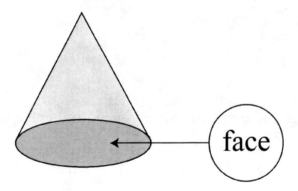

A **cube** has **6 faces**.
All 6 faces are shaped like squares.

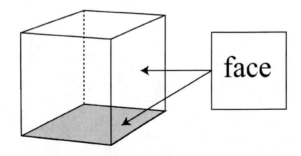

Naming Faces of Solid Figures

A **rectangular prism** has **6 faces**.

The faces are shaped like rectangles. Two of the faces of a rectangular prism can be shaped like squares.

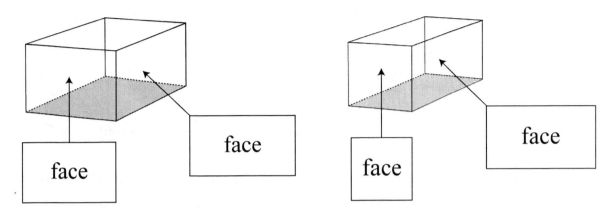

Look at the object. Circle all the shapes that make up the object.

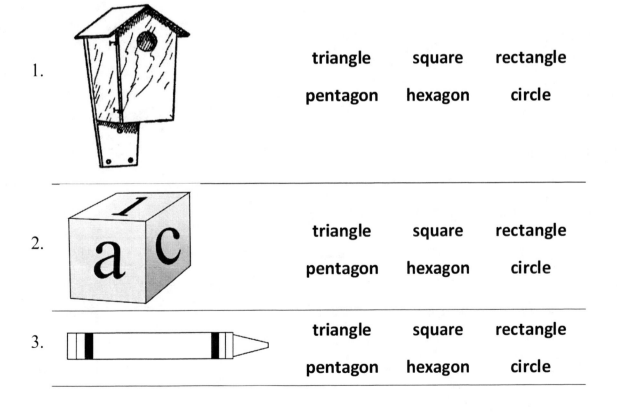

1. triangle   square   rectangle
   pentagon   hexagon   circle

2. triangle   square   rectangle
   pentagon   hexagon   circle

3. triangle   square   rectangle
   pentagon   hexagon   circle

Chapter 10   Geometry

Compare Shapes

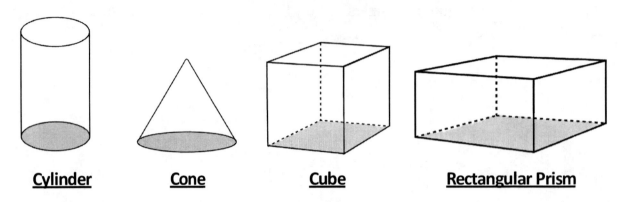

Cylinder   Cone   Cube   Rectangular Prism

Look at the shapes and solids above. Circle yes or no if they can roll, slide or stack.

(Hint: you can put some of the objects on their side to make them roll.)

Object	Roll?	Slide?	Stack?
1. Cylinder	yes no	yes no	yes no
2. Cone	yes no	yes no	yes no
3. Cube	yes no	yes no	yes no
4. Rectangular Prism	yes no	yes no	yes no

Read the question. Circle yes or no for your answer.

5. Can you make a round ball out of a rectangular prism?   yes   no

6. Can you make a tube out of a cylinder?   yes   no

7. Can you make a box out of a cone?   yes   no

8. Can you make a box out of a cube?   yes   no

Create Pictures Using Shapes

Answer the questions below using the picture of the house.

1. What shape are the black shutters? _____

2. What shape are the columns on the porch? _____

3. What shape are the steps? _____

4. The windows are rectangles divided into what shape? _____

5. What shape is the long hedge on the front of the porch? _____

Color the house. Squares are blue, rectangles are green, and cylinders are yellow.
Color the rest of the shapes any color you want.

Chapter 10   Geometry

Plane Figures

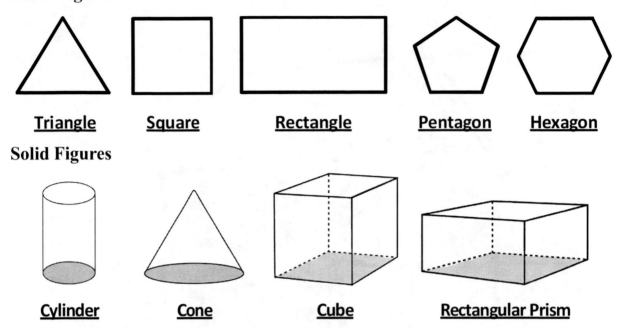

Use the plane figures and the solid figures above to make a picture.
Use at least 2 plane figures and 2 solid figures.
The rest of the drawing can be filled in as you wish.
Color the picture any colors you want.

Overlapping Shapes

EXAMPLE: What happens if you **overlap** a triangle onto a rectangle?

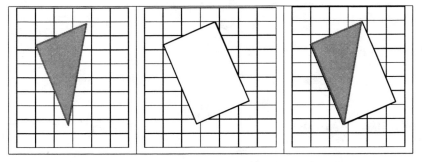

You get 2 triangles.

EXAMPLE: What happens if you **overlap** a pentagon onto a square?

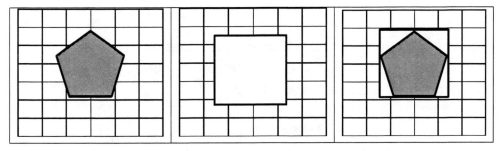

You get a pentagon surrounded by 4 triangles.

You are given 2 shapes.
Make new shapes with them by overlapping the 2 shapes.
Have fun with this. Color the shapes when you are done.

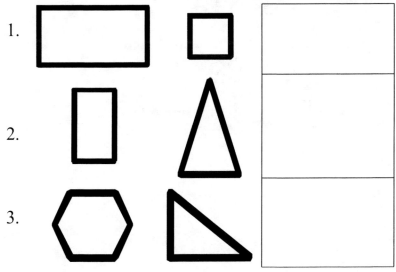

Chapter 10 Geometry

Position

Position is <u>where</u> one object is next to another object.

EXAMPLE: The acorn is **above** the apple.

The apple is **below** the acorn.

EXAMPLE: **Look at the penguin picture below.**
  a) The 1st penguin is **in front of** the 2nd and 3rd penguins.
  b) The 2nd and 3rd penguins are **next to** each other.
  c) The 4th penguin is **in back of** the 2nd and 3rd penguins.
  d) The last penguin is **far** away.
  e) The 1st penguin is **near**.
  f) The 2nd penguin is **to the left of** the 3rd penguin.
  g) The 3rd penguin is **to the right of** the 2nd penguin.
  h) The last penguin is **behind** all of the other penguins.

Position

Look at the picture above.
Read the question. Circle the correct answer.
Color the picture.

1.	Where is the horse?	
	in front of the wagon	on top of the haystacks
2.	Where is the man carrying 2 pumpkins?	
	to the left of the horse	in back of the wagon
3.	Where are all the haystacks?	
	in front of the wagon	far away
4.	Where are the pumpkins?	
	in back of the wagon, in front of the wagon, to the right and left of the wagon, near, and far	the men are carrying the only pumpkins

Chapter 10   Geometry

Direction

Directions describe which way. We will use **up**, **down**, **left**, and **right**.

EXAMPLE: Look at the picture below.
                The chick is looking **up** at the chicken.
                The chicken is looking **down** on her chick.

EXAMPLE: Look at the picture below.
                The mama dog is looking to the **left**.
                The papa dog is looking to the **right**.

Answer the questions below using the words <u>up</u>, <u>down</u>, <u>left</u>, or <u>right</u>.

1. Where is the ceiling? _____

2. Where is the floor? _____

3. Is this a left hand or a right hand? _____

4. Is the kangaroo facing left or right? _____

Chapter 10 Review

Write the name of each of the plane figures.
Choose from these names: triangle   square   rectangle   pentagon   hexagon
Next, tell how many sides and corners each plane figure has.

Plane Figure	Name	Number of Sides	Number of Corners
1. ⬠	_____	_____	_____
2. ☐	_____	_____	_____
3. ⬡	_____	_____	_____
4. △	_____	_____	_____
5. ▭	_____	_____	_____

Look at the solid figure. Count how many faces and corners. Write the number.

Solid Figure	Name	Number of Faces	Number of Corners
6. cylinder	_____	_____	_____
7. cone	_____	_____	_____
8. cube	_____	_____	_____
9. rectangular prism	_____	_____	_____

Chapter 10   Geometry

Name the shape of the face on each figure.

Figure	Name the Face
10.	_____
11.	_____
12.	_____

Use the picture of the baseball and glove to answer the questions.
Fill in the blanks with either <u>in front of</u> or <u>behind</u>.

13.   The baseball is _____ the glove.

14.   The glove is _____ the baseball.

Use the picture of the mouse to answer the question.
Fill in the blank with either <u>up</u>, <u>down</u>, <u>left</u>, or <u>right</u>.

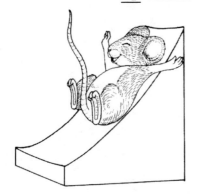

15. The mouse is sliding _____ the slide.

Chapter 10 Test

Read the directions and circle the right answer.

1. What is the name of the plane figure below?

    A square

    B pentagon

    C hexagon

2. What is the name of the solid figure below?

    A cube

    B cone

    C cylinder

3. What is the name of the solid figure that is most like the party hat below?

    A cube

    B cone

    C cylinder

4. Which plane figure has 3 sides and 3 corners?

    A hexagon

    B square

    C triangle

Chapter 10   Geometry

Use the picture of fruit to fill in the blanks for questions on this page.

5. The bananas are _____ the apple.

   A in back of

   B in front of

   C far away from

6. The cherries are _____ the strawberry.

   A to the left of

   B to the right of

   C in front of

7. The bananas are _____ the pear.

   A in back of

   B in front of

   C on top of

Chapter 10 Test

8. If you are outside, where is the sky?

   A to your left

   B below

   C above

9. Fill in the blank: The penny is to the _____ of the nickel.

   A right

   B left

   C below

10. What is the name of the figure below?

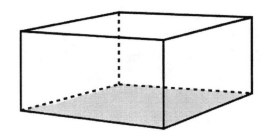

   A cube

   B rectangular prism

   C cone

Chapter 11
Graphs

This chapter covers the following Georgia Performance Standards:

M1D	Data Analysis and Probability	M1D1a, b

Tables

A **table** is used to put facts in order.

EXAMPLE: Below is a table of favorite ice cream of 1st graders at Hope Elementary School.

	Chocolate	Vanilla	Strawberry
Boys	50	40	35
Girls	45	30	50
Teachers	2	1	1

Look at the table. The kinds of ice cream go across the top row.
The kinds of people are in the first column.
The table is filled in with boys', girls' and teachers' favorite ice cream.

Question: How many boys like vanilla ice cream?
First, go to the line for boys.
Next, go across to the column for vanilla.
The number of boys who like vanilla is 40.

Question: How many girls like strawberry ice cream?
First, go to the line for girls.
Next, go across to the column for strawberry.
The number of girls who like strawberry is 50.

Read each table. Answer the questions after each table. Write the answer.

Allgood Elementary School: 1st Grade
Favorite Fruit

	Bananas	Apples	Grapes
Boys	25	30	20
Girls	15	40	20
Teachers	1	2	1

1. How many teachers like apples? _____

2. How many girls like grapes? _____

3. How many boys like bananas? _____

4. How many girls like bananas? _____

Blue Water Elementary School: 1st Grade
Favorite Outdoor Game

	Soccer	Hopscotch	Playing Catch
Boys	40	0	35
Girls	30	25	15
Teachers	2	1	1

5. How many boys like soccer? _____

6. How many girls like hopscotch? _____

7. How many teachers like playing catch? _____

8. How many boys like playing catch? _____

Chapter 11 Graphs

Tally Charts

Tally Charts are used to sort facts.

EXAMPLE: Below is a list of names and each person's favorite color.

Name	Favorite Color
Jean	blue
John	red
Kari	green
Alice	pink
Kirk	blue
Mary	pink
Dan	blue
Mario	red
Sara	green
Rick	blue
Ana	pink

To find how many people like each color, make a tally chart.
Write a list of color names.
Make a **tally** mark by the color for each person's favorite.

Favorite Color	Tally
blue	/ / / /
green	/ /
pink	/ / /
red	/ /

Question: How many people like green? Kari and Sara.
2 people like green.

Question: How many people like blue? Jean, Kirk, Dan, and Rick.
4 people like blue.

Tally Charts

Read the list of people and their favorite vegetable.
Make a tally chart.

Name	Favorite Vegetable
Jake	carrot
Ching	cabbage
Mario	tomato
Dave	tomato
Alice	carrot
Robert	cabbage
Amy	lettuce
Jerry	carrot
Anne	carrot

Tally Chart:

Favorite Vegetable	Tally Marks
carrot	
cabbage	
lettuce	
tomato	

Read the list of people and their favorite subject in school.
Make a tally chart.

Name	Favorite Subject
Max	math
Ira	science
Della	math
Rita	math
Bobby	gym
Emma	art
Jim	science
Kim	reading

Tally Chart:

Favorite Subject	Tally Marks
math	
reading	
gym	
science	
art	

Chapter 11   Graphs

Making Tables and Tally Charts

Make a table of favorite gum to chew.
Ask 8 kids in class what their favorite gum is.
Write the names in the first column.
Write their favorite gum flavor in the second column.

Name	Favorite gum

Make a tally chart from the facts you collected above.
Write each favorite gum flavor in the first column.
Put one tally mark for each person who liked that flavor of gum best.

Favorite Gum	Tally Marks

Picture Graphs

Graphs are used to display facts.

Picture graphs use pictures to stand for certain amounts.

Picture graphs use a **key** to show how many each picture stands for.

EXAMPLE: The picture graph below shows the favorite treats of a group of kids.

Favorite Treat	Number of Kids
Ice Cream	👤 👤
Cupcakes	👤 👤 👤
Apples	👤

Key: 👤 = 5 kids

The key says each  👤 = 5 kids, so for every 👤 , count by 5's.

How many kids like ice cream best? 5, 10. 10 kids like ice cream best.

How many kids like cupcakes best? 5, 10, 15. 15 kids like cupcakes best.

How many kids like apples best? 5. 5 kids like apples best.

Go back up to the picture graph. Color the stick people.
Use orange for ice cream, blue for cupcakes, and yellow for apples.

Chapter 11   Graphs

Make your own pictograph. Use the chart for your facts.
Use a circle and color it in for your pictures. (You may use any color you want.)
Each circle = 10 people.

FACT CHART

Favorite Kind of Movie	Number of People
Cartoon	40
Animal	30
Funny	30
Musical	20

PICTURE GRAPH

KEY:
◯ = 10 people

Favorite Kind of Movie	Number of People
Cartoon	
Animal	
Funny	
Musical	

Bar Graphs

Bar graphs are another way of showing the facts.

The bar graph shows the names on one side and the numbers on the side next to it.

EXAMPLE: The bar graph below shows the same favorite treats you saw in the picture graph section.
Instead of using pictures, we will use bars to show the facts.

How many people like ice cream? Find the ice cream column and go up until you reach the top of the gray box. 10 is at the top. 10 people like ice cream.

Do the same for the cupcakes and apples.

Make a bar graph using the favorite movie facts from the last section in this book.
Use any colors you want.
Cartoon = 40   Animal = 30   Funny = 30   Musical = 20

Chapter 11   Graphs

Making Bar Graphs

The first box is a tally chart. Read it.
Then, make your own bar graph using the facts from the tally chart.
You may use any colors you want.

Favorite Game	Tally
Tag	//
Soccer	卌 //
Baseball	卌
Board Games	////

BAR GRAPH:

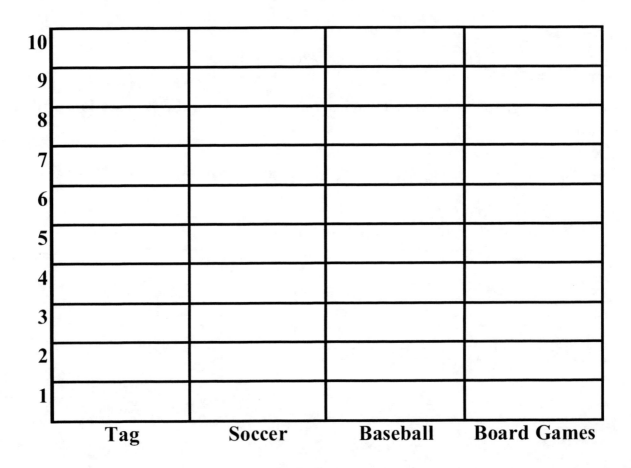

Chapter 11 Review

Look at the table and answer the questions.

BRING A BUG TO CLASS DAY

	Ant	Fly	Beetle
Boys	10	15	20
Girls	5	30	10
Teachers	1	2	1

1. How many boys brought beetles to class? _____

2. How many teachers brought flies to class? _____

3. How many girls brought ants to class? _____

4. How many girls brought flies to class? _____

Use the list to make a tally chart of favorite colors.

Name	Favorite Color
Brandon	blue
Clare	red
Kim	pink
Ted	blue
Mary	green
Jose	blue
Edna	pink
Jake	red
Andre	blue

	Favorite Color	Tally
5.	blue	
6.	red	
7.	pink	
8.	green	

Chapter 11   Graphs

Read the picture graph below and answer the questions.

Key:

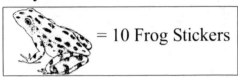
= 10 Frog Stickers

Name	Number of Frog Stickers
Henry	🐸 🐸 🐸
John	🐸 🐸 🐸 🐸
Eddie	🐸 🐸 🐸

9. How many Frog Stickers does Henry have? _____

10. Which two boys have the same number of frog stickers? _____

11. How many frog stickers does John have? _____

12. How many frog stickers does Eddie have? _____

13. How many frog stickers do all 3 boys have? (Add all 3) _____

14. How many more frog stickers does John have than Eddie? _____

15. Add together the frog stickers Henry and Eddie have. _____

Color the frogs if you have time.

Chapter 11 Test

Look at the table below and circle the right answer for each question.

Favorite Flowers

	daisy	rose	pansy
Boys	10	20	5
Girls	15	15	10
Teachers	1	3	1

1. How many teachers like this flower best?

   A 1
   B 2
   C 3

2. How many boys like this flower best?

   A 10
   B 20
   C 30

3. How many girls like this flower best?

   A 10
   B 20
   C 30

4. How many girls like this flower best?

   A 10
   B 15
   C 20

Chapter 11   Graphs

Squirrel	Number of Acorns Hidden
(squirrel 1)	🌰🌰🌰🌰🌰
(squirrel 2)	🌰🌰🌰🌰
(squirrel 3)	🌰🌰🌰🌰🌰🌰🌰

Key:
🌰 = 10

5. How many did  hide?

   A 30

   B 40

   C 50

6. How many did hide?

   A 30

   B 40

   C 50

7. How many did hide?

   A 30

   B 50

   C 70

Chapter 12
Addition and Subtraction with 2-Digit Numbers

This chapter covers the following Georgia Performance Standards:

M1N	Number and Operations	M1N3f
		M1N3g, h

Add Tens

You know how to count by tens. **Adding by tens** is just as easy.

EXAMPLE: 20   ⇨   Think: 2 tens   ⇨   20
             + 50   + 5 tens   + 50
               ?   7 tens   ⇨ 7 tens = 70   70

Add. Think tens. Write your answer.

1. 10   3. 30   5. 20   7. 60   9. 10
   + 30   + 40   + 20   + 10   + 80
    40

2. 30   4. 40   6. 30   8. 20   10. 50
   + 20   + 50   + 30   + 10   + 30

Chapter 12   Addition and Subtraction with 2-Digit Numbers

More practice adding tens.

Add. Think tens. Write your answer.

11. A bear ate 40 blueberries. Then it ate 20 more. How many blueberries did the bear eat in all?

    40 + 20 = _____

12. The next day the bear ate 30 blueberries. Then it ate 10 more. How many blueberries did the bear eat in all?

    30 + 10 = _____

13. Jamie's mom made 20 cupcakes. Then she made 30 more.

    20 cupcakes + 30 cupcakes = _____

14. The next day, Jamie's mom made 40 cupcakes. Then she made 40 more.

    40 cupcakes + 40 cupcakes = _____

Add Tens and Ones

EXAMPLE: Add 20 + 4.   20   2 tens
           Think 2 tens + 4 ones.   + 4   4 ones

                                     24

tens	ones
2	0
+	4
2	4

 +   = 24¢

Add the tens and ones. Write your answer.

1. 70   3. 30   5. 80   7. 90   9. 20
   + 2   + 1   + 3   + 7   + 6

    72

2. 50   4. 60   6. 20   8. 10   10. 70
   + 4   + 8   + 1   + 5   + 9

Add 2-Digit Numbers

When adding **2-digit numbers**, <u>first</u> add the <u>ones</u>, then add the tens.

EXAMPLE:

tens	ones
1	2
2	4

12
+ 24

First:

ones
2
4
6

Next:

tens	ones
1	2
2	4
3	6

 = 36¢

Add. Write your answer.

1. 42
   + 13
    55

2. 63
   + 16

3. 13
   + 43

4. 17
   + 32

5. 72
   + 26

6. 62
   + 27

7. 50
   + 18

8. 43
   + 56

9. 51
   + 12

10. 33
   + 11

11. 73
   + 25

12. 17
   + 31

13. 24
   + 15

14. 12
   + 10

15. 19
   + 60

Chapter 12   Addition and Subtraction with 2-Digit Numbers

Subtract Tens

You know how to subtract by ones. **Subtracting by tens** is the next step.

EXAMPLE: $\begin{array}{r}40\\-10\\\hline ?\end{array}$ ⇒ Think: $\begin{array}{r}4\text{ tens}\\-1\text{ ten}\\\hline 3\text{ tens}\end{array}$ ⇒ 3 tens = 30 ⇒ $\begin{array}{r}40\\-10\\\hline 30\end{array}$

🪙🪙🪙🪙 − 🪙 = 30¢

Subtract. Think tens. Write your answer.

1. $\begin{array}{r}80\\-20\\\hline 60\end{array}$ 3. $\begin{array}{r}30\\-10\\\hline\end{array}$ 5. $\begin{array}{r}70\\-40\\\hline\end{array}$ 7. $\begin{array}{r}90\\-30\\\hline\end{array}$ 9. $\begin{array}{r}90\\-70\\\hline\end{array}$

2. $\begin{array}{r}40\\-20\\\hline\end{array}$ 4. $\begin{array}{r}20\\-10\\\hline\end{array}$ 6. $\begin{array}{r}30\\-30\\\hline\end{array}$ 8. $\begin{array}{r}80\\-40\\\hline\end{array}$ 10. $\begin{array}{r}70\\-20\\\hline\end{array}$

11. 🪙🪙🪙🪙 − 🪙🪙🪙 = ___¢

12. 🪙🪙🪙🪙🪙 − 🪙🪙 = ___¢

13. 🪙🪙🪙🪙🪙🪙 − 🪙🪙 = ___¢

14. 🪙🪙🪙🪙🪙🪙🪙🪙 − 🪙 = ___¢

15. Josh threw the ball 40 times. He missed 20 times.
How many baskets did Josh make? 40 – 20 = _____

16. The next day Josh threw the ball 50 times. He missed 20.
How many baskets did he make the next day? 50 – 20 = _____

17. Farmer Brown had a herd of 90 cows. He took
30 cows to market to sell. How many cows
does Farmer Brown have left? 90 – 30 = _____

18. The next month, Farmer Brown took 20 more
cows to market to sell. How many cows does
Farmer Brown have now? 60 – 20 = _____

Subtract Tens and Ones

EXAMPLE: Subtract: 36 – 5.

```
  36
 – 5
  31
```

tens	ones
3	6
	5
3	1

Subtract the tens and ones. Write your answer.

1. 55
   – 4
    51

2. 10
   – 7

3. 69
   – 9

4. 88
   – 6

5. 99
   – 3

6. 11
   – 8

7. 45
   – 2

8. 53
   – 3

9. 87
   – 4

10. 76
   – 4

Chapter 12 Addition and Subtraction with 2-Digit Numbers

Subtract 2-Digit Numbers

When subtracting **2-digit numbers**, first subtract the ones, then subtract the tens.

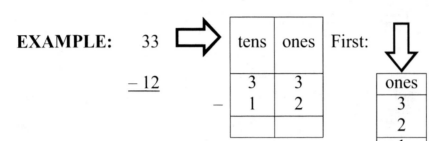
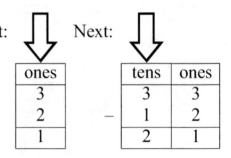

Subtract. Write your answer.

1. 46
  − 11
   35

2. 67
  − 32

3. 93
  − 51

4. 85
  − 12

5. 59
  − 18

6. 89
  − 47

7. 29
  − 14

8. 48
  − 27

9. 23
  − 12

10. 54
   − 33

11. 22
   − 12

12. 56
   − 33

13. 72
   − 51

14. 49
   − 37

15. 46
   − 20

Mixed Practice

Add or subtract. Write your answer.

1. $\begin{array}{r} 21 \\ +7 \\ \hline 28 \end{array}$

2. $\begin{array}{r} 60 \\ +30 \\ \hline \end{array}$

3. $\begin{array}{r} 78 \\ -47 \\ \hline \end{array}$

4. $\begin{array}{r} 22 \\ -11 \\ \hline \end{array}$

5. $\begin{array}{r} 88 \\ -35 \\ \hline \end{array}$

6. $\begin{array}{r} 17 \\ -5 \\ \hline \end{array}$

7. $\begin{array}{r} 92 \\ -61 \\ \hline \end{array}$

8. $\begin{array}{r} 54 \\ -22 \\ \hline \end{array}$

9. $\begin{array}{r} 40 \\ +50 \\ \hline \end{array}$

10. $\begin{array}{r} 90 \\ -30 \\ \hline \end{array}$

11. $\begin{array}{r} 20 \\ +20 \\ \hline \end{array}$

12. $\begin{array}{r} 47 \\ +31 \\ \hline \end{array}$

13. $\begin{array}{r} 35 \\ +23 \\ \hline \end{array}$

14. $\begin{array}{r} 83 \\ +5 \\ \hline \end{array}$

15. $\begin{array}{r} 28 \\ -14 \\ \hline \end{array}$

16. $\begin{array}{r} 14 \\ +44 \\ \hline \end{array}$

17. $\begin{array}{r} 36 \\ -24 \\ \hline \end{array}$

18. $\begin{array}{r} 88 \\ -22 \\ \hline \end{array}$

19. $\begin{array}{r} 60 \\ -30 \\ \hline \end{array}$

20. $\begin{array}{r} 40 \\ -10 \\ \hline \end{array}$

21. $\begin{array}{r} 62 \\ +36 \\ \hline \end{array}$

22. $\begin{array}{r} 75 \\ -25 \\ \hline \end{array}$

23. $\begin{array}{r} 17 \\ +50 \\ \hline \end{array}$

24. $\begin{array}{r} 15 \\ +14 \\ \hline \end{array}$

25. $\begin{array}{r} 81 \\ +12 \\ \hline \end{array}$

26. $\begin{array}{r} 94 \\ -30 \\ \hline \end{array}$

27. $\begin{array}{r} 50 \\ -30 \\ \hline \end{array}$

28. $\begin{array}{r} 60 \\ -10 \\ \hline \end{array}$

29. $\begin{array}{r} 62 \\ -11 \\ \hline \end{array}$

30. $\begin{array}{r} 70 \\ -20 \\ \hline \end{array}$

Chapter 12   Addition and Subtraction with 2-Digit Numbers

Adding Money

EXAMPLE:

Add.
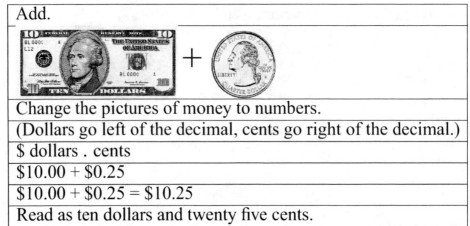
Change the pictures of money to numbers.
(Dollars go left of the decimal, cents go right of the decimal.)
$ dollars . cents
$10.00 + $0.25
$10.00 + $0.25 = $10.25
Read as ten dollars and twenty five cents.

Add the money problems. Write your answer.
Don't forget the $ and decimal.

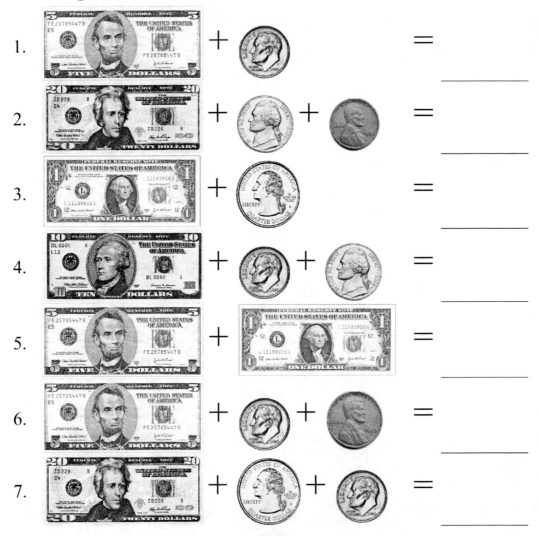

Subtracting Money

EXAMPLE:

Subtract.
$5 bill − $1 bill
Change the pictures of money to numbers.
(Dollars go left of the decimal, cents go right of the decimal.)
$ dollars . cents
$5.00 − $1.00
$5.00 − $1.00 = $4.00
Read as four dollars.

Add the money problems. Write your answer.
Don't forget the $ and decimal.

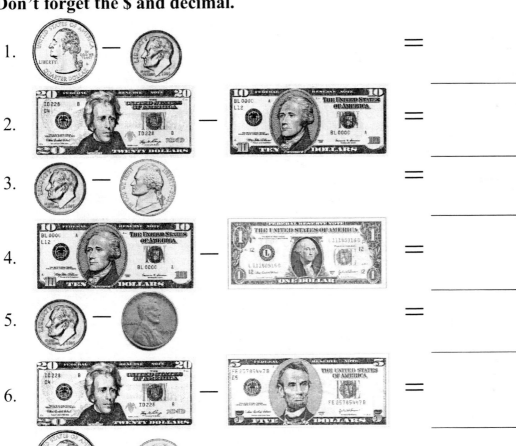

1. quarter − dime = _____
2. $20 − $10 = _____
3. dime − nickel = _____
4. $10 − $1 = _____
5. dime − penny = _____
6. $20 − $5 = _____
7. quarter − nickel = _____

Chapter 12   Addition and Subtraction with 2-Digit Numbers

Choose the Operation

Choose plus or minus to fit the problem.

EXAMPLE: 6 ☐ 2 = 4

Which sign, + or –, fills in the ☐ correctly?
Try +.   6 + 2 = 8   No, + does not fit.
Try –.   6 – 2 = 4   Yes, – does fit. 6 – 2 is equal to 4.

Read each problem. Which sign fills in the ☐ ?
Circle the correct sign.

                                Plus   Minus

1. 8 ☐ 2 = 10   +   –

2. 20 ☐ 5 = 15   +   –

3. 12 ☐ 4 = 16   +   –

4. 90 ☐ 30 = 60   +   –

5. 75 ☐ 5 = 80   +   –

6. 52 ☐ 6 = 58   +   –

7. 20 ☐ 8 = 28   +   –

8. 9 ☐ 4 = 5   +   –

9. 33 ☐ 3 = 36   +   –

10. 15 ☐ 3 = 12   +   –

11. 48 ☐ 26 = 22   +   –

12. 99 ☐ 9 = 90   +   –

Word Problems

Read the word problems. Write your answer.

1. Toby and his family went to the zoo.
   They saw 2 alligators in a pit and 3 more out of the pit.
   How many alligators did they see in all? _____

2. An alligator ate 2 fish. The next day the alligator ate 4 fish.
   How many fish did the alligator eat over the two days? _____

3. Amy colored 2 pages on Monday.
   She colored 3 more pages on Tuesday.
   How many pages did Amy color over the two days? _____

4. Amy had 17 crayons. She gave her brother 6
   crayons. How many crayons did Amy have left? _____

5. A hippo bellowed 3 times in the morning.
   He bellowed 4 times in the afternoon.
   How many times did the hippo bellow? _____

6. During 12 hours, the hippo at the zoo spent 7
   hours <u>in</u> the water. How many hours did it spend
   <u>out</u> of the water in the same 12 hours? _____

7. The tiger at the zoo had 2 cubs this year.
   Last year she had 4 cubs. How many cubs
   did the tiger have over the two years? _____

8. A tiger in the wild ran after a group of
   16 antelope. 12 of the antelopes ran away.
   How many antelopes stayed still? _____

9. A dolphin swam 2 miles in the morning.
   It swam 3 miles in the afternoon.
   How many miles did it swim in all? _____

Chapter 12 Addition and Subtraction with 2-Digit Numbers

Chapter 12 Review

Add. Write your answer.

1. 25
  + 3

2. 40
  +20

3. 95
  + 3

4. 23
  +12

5. 80
  +14

Subtract. Write your answer.

6. 68
  − 7

7. 38
  −22

8. 90
  −80

9. 18
  − 5

10. 72
  −31

Add or subtract the money problems. Write your answer. Don't forget the $ and decimal.

11. $1 + nickel = _____

12. $5 − $1 = _____

13. $20 + quarter + dime = _____

Read each problem. Which sign fills in the ☐ ? Circle the correct sign.

Plus   Minus

14. 7 ☐ 3 = 4   +   −

15. 68 ☐ 5 = 63   +   −

16. 20 ☐ 8 = 28   +   −

210   Copyright © American Book Company

Chapter 12 Test

Read the directions and circle the right answer.

1. Add. $\begin{array}{r} 34 \\ + 15 \\ \hline \end{array}$

    A 39

    B 49

    C 57

2. Subtract. $\begin{array}{r} 87 \\ - 55 \\ \hline \end{array}$

    A 32

    B 42

    C 52

3.  $5 bill + dime + nickel =

    A $5.00

    B $5.10

    C $5.15

4.  $20 bill − $1 bill =

    A $19.00

    B $20.00

    C $21.00

5.  4 dimes − 3 dimes =

    A 10 ¢

    B 20 ¢

    C 30 ¢

6. Harry, the raccoon, caught 2 fish in the morning.
He caught 3 more fish in the afternoon.
How many fish did Harry catch in all?

A 2

B 3

C 5

7. Which sign fills in the ☐ ?

36 ☐ 24 = 12

A +

B −

C =

8.

A 10 ¢

B 20 ¢

C 30 ¢

Practice Test 1

Part 1

Read the directions and circle the right answer.

1. What is sixteen in number form?

   A 16
   B 60
   C 61

   M1N1a

2. Count how many.

   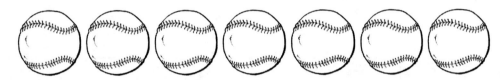

   A 5
   B 6
   C 7

   M1N1b

3. Which number sentence is correct?

   A 2 < 8
   B 8 < 2
   C 2 > 8

   M1N1c

4. Which is a fair trade?

   A

   B

   C

   M1N1e

5. What is the value of the bill below?

A $1.00
B $10.00
C $50.00

M1N1f

6. Look at the numbers chart below. Which number is 38 closest to?

31	32	33	34	35	36	37	38	39	40
41	42	43	44	45	46	47	48	49	50
51	52	53	54	55	56	57	58	59	60

A 40
B 50
C 60

M1N2a

7. How many groups of tens?

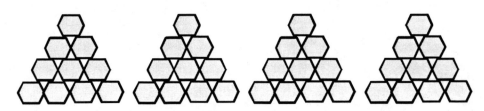

A 4 groups of tens
B 5 groups of tens
C 6 groups of tens

M1N2b

8. How many tens and ones does 91 have?

A | Tens | Ones |
  | --- | --- |
  | 9 | 1 |

B | Tens | Ones |
  | --- | --- |
  | 1 | 9 |

C | Tens | Ones |
  | --- | --- |
  | 9 | 9 |

9. What is ten less than 83?

   A 93
   B 83
   C 73

10. Which number sentence is equal to 6?

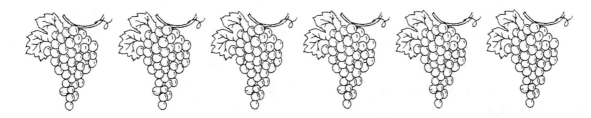

   A 3 + 3 = 6
   B 4 + 3 = 6
   C 3 + 2 = 6

11. You have 4 peanuts. You count on 2 more. How many peanuts do you have now?

   A 4
   B 5
   C 6

12. Add. $\begin{array}{r}53\\+46\\\hline\end{array}$

   A 99
   B 88
   C 77

M1N3g

13. A snail crawled 3 inches. Then the snail crawled 7 more inches. How many inches did the snail crawl in all?

   A 7
   B 10
   C 12

M1N3h

14. Mario's cat had 6 kittens. He gave 4 of the kittens away. How many kittens did Mario keep?

   A 6
   B 4
   C 2

M1N3h

15. Emma, Lu, and Marie played hopscotch.
    They <u>each</u> took 10 turns.
    How many turns did the 3 girls take in all?

    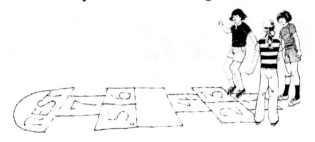

    A 60
    B 30
    C 10

16. There was a whole sandwich on the table.
    Mom cut the sandwich in 2 parts.
    How did she cut the sandwich?

    A in halves
    B in fourths
    C one whole

17. Which weighs the least? A baseball, a balloon, or a horse?

    A baseball
    B balloon
    C horse

18. About how many paper clips long is the rope?

   A 2 paper clips long
   B 4 paper clips long
   C 6 paper clips long

19. About how many thumb widths is the fish?
    (Use the thumb widths in the drawing. Not your own thumb width.)

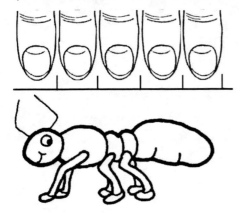

   A 3
   B 4
   C 5

20. What time is it to the nearest hour?

   A 1:00
   B 8:00
   C 9:00

21. Which takes the longest to do?

   A tie your shoes
   B read for 2 hours
   C do 50 jumping jacks

22. How many corners does a pentagon have?

   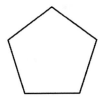

   A 4
   B 5
   C 6

23. How many sides does a square have?

   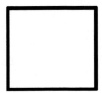

   A 4
   B 5
   C 6

24. What is the name of the object below?

   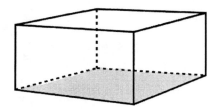

   A cube
   B rectangular prism
   C cone

25. Which solid is this toy most like?

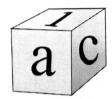

   A cube
   B cylinder
   C cone

M1G1b

26. Which of these shapes has 4 sides and 4 corners?

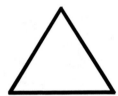  

   A triangle
   B square
   C hexagon

M1G2

27. Where is the worm's head?

   A The worm's head is on top of the apple.
   B The worm's head is in back of the apple.
   C The worm's head is in front of the apple.

M1G3

220 Copyright ©American Book Company

Use this chart for the next 3 problems.

Favorite Things to do Outside

	Bicycling	Basketball	Catch Bugs
	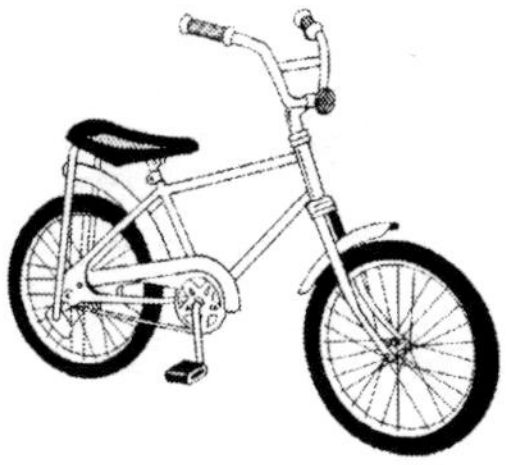	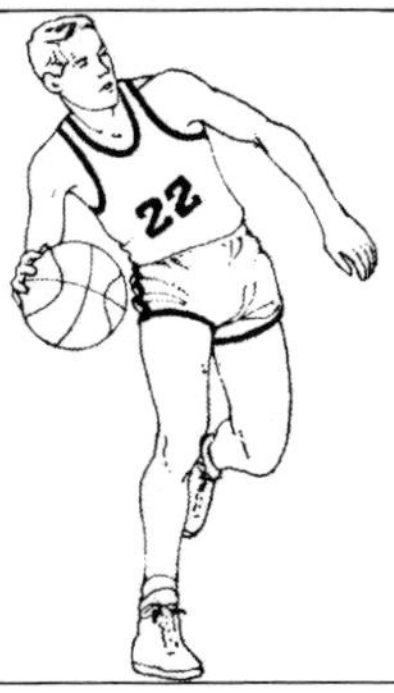	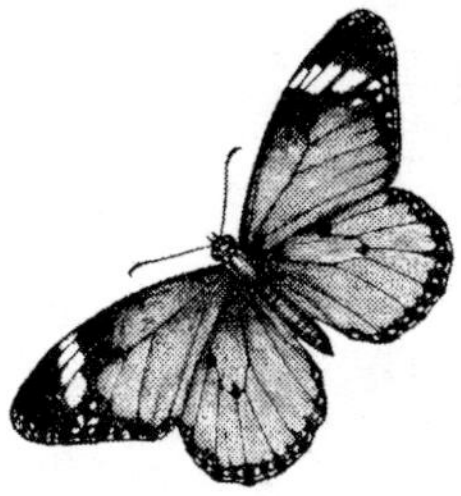
Boys	4	4	2
Girls	4	1	1

28. How many girls like to bicycle?

A 1
B 2
C 4

M1D1a

29. How many boys like to play basketball?

A 1
B 2
C 4

M1D1a

30. Which tally chart shows the favorite things the girls like to do?

Chart 1	Count
Bicycling	////
Basketball	/
Catch Bugs	////

Chart 2	Count
Bicycling	////
Basketball	//
Catch Bugs	//

Chart 3	Count
Bicycling	////
Basketball	/
Catch Bugs	/

A Chart 1

B Chart 2

C Chart 3

M1D1b

Part 2

31. The number 12 is written as ____ .

   A ten
   B eleven
   C twelve

32. Count how many.

   A 4
   B 5
   C 6

33. Which number sentence is correct?

   A 19 < 18
   B 18 < 19
   C 18 > 19

34. Look at the number line. What number is missing?

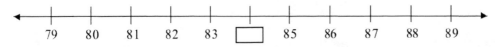

   A 80
   B 82
   C 84

35. Lisa has 60 ¢. She wants to buy a cupcake that costs 50 ¢.
Does Lisa have enough money? Will she have money leftover?

 — 50 ¢

A Yes, and she will have 10 ¢ leftover.
B No, she doesn't have enough money.
C Yes, but she will have no money leftover.

M1N1e

36. Look at the number below.
Which number is it closest to?

19

A 0
B 10
C 20

M1N2a

37. How many tens and ones does 24 have?

A
Tens	Ones
2	4

B
Tens	Ones
4	2

C
Tens	Ones
24	0

M1N2c

38. Skip count by 5's. What is the missing number?

5, 10, ___ , 20, 25

A 11
B 15
C 19

M1N3b

39. Which number sentence is equal to 5?

   A $1 + 5 = 5$
   B $1 + 4 = 5$
   C $4 + 2 = 5$

M1N3c

40. Count how many bunnies are in the 1st set.
    Count how many bunnies are in the 2nd set.
    What is Set 1 minus Set 2?

   A 2
   B 3
   C 5

M1N3d

41. Start at 72. Count forward 4.
    What is the new number?

   A 73
   B 75
   C 76

M1N3e

42. Which sentence is true?

   A $6 + 4 = 4 + 6$
   B $6 + 4 = 64$
   C $6 + 4 = 9$

M1N3f

43. $\begin{array}{r}70\\-40\\\hline\end{array}$

  A 70
  B 40
  C 30

44. Matt had 12 bananas.
    He shared 8 bananas with friends.
    How many bananas does Matt have left?

  A 6
  B 4
  C 2

45. Lu Ann was matching socks from the dryer.
    She found a total of 9 socks.
    How many pairs of socks can she match up? Will there be any leftover?

  A She can make 5 pairs with 1 sock leftover.
  B She can match 4 pairs with 1 sock leftover.
  C She can match 4 pairs with no socks leftover.

46. Is the number 12 even or odd?

  A even
  B odd
  C both

47. How many parts is the rectangle divided into?

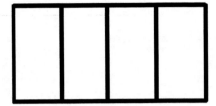

   A 2
   B 3
   C 4

MIN4c

48. Which weighs the least? A hamster, a butterfly, or a camel?

   A hamster
   B butterfly
   C camel

M1M1a

49. About how many paper clips long is the scissors?

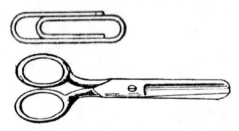

   A 1 paper clip long
   B 2 paper clips long
   C 3 paper clips long

M1M1b

50. What order are the months of the year?

A

January	February	March	May	April	June
July	August	September	November	October	December

B

January	February	March	May	April	July
June	August	September	October	November	December

C

January	February	March	April	May	June
July	August	September	October	November	December

51. Which takes the shortest amount of time to do?

A Spend the day at the zoo.

B Read 15 pages from a book.

C Clap your hands 15 times.

52. Which shape is a dollar bill most like?

A square

B rectangle

C triangle

53. How many sides does a hexagon have?

Hexagon

A 4
B 6
C 8

M1G1a

54. What is the name of the object below?

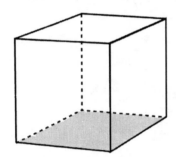

A cube
B cylinder
C cone

M1G1b

55. Which solid is the crown most like?

A cube
B cylinder
C cone

M1G1b

56. Which of these shapes has 5 sides and 5 corners?

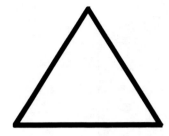  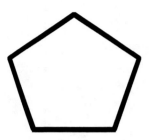

A triangle

B square

C pentagon

M1G2

57. Where is the compass?

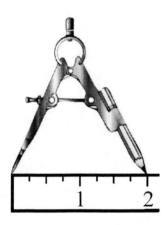

A The compass is under the ruler.

B The compass is above the ruler.

C The compass is behind the ruler.

M1G3

Use this chart for the next 3 problems.

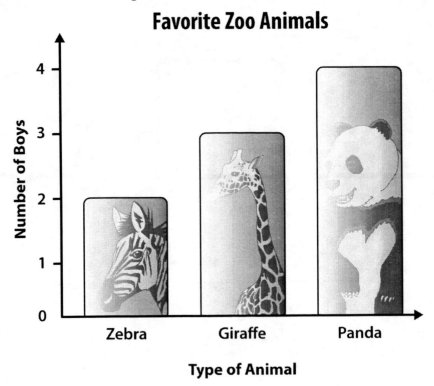

58. How many boys like the pandas best?

   A 4
   B 3
   C 2

M1D1a

59. How many boys like giraffes best?

   A 4
   B 3
   C 2

M1D1a

60. Which tally chart shows the favorite zoo animals of boys?

Chart 1	Count
Zebra	/ /
Giraffe	/ / /
Panda	/ / / /

Chart 2	Count
Zebra	/ / /
Giraffe	/ / / /
Panda	/ /

Chart 3	Count
Zebra	/ / /
Giraffe	/ /
Panda	/ / / /

   A Chart 1
   B Chart 2
   C Chart 3

M1D1b

Practice Test 2

Part 1

Read the directions and circle the right answer.

1. What is 41 in word form?

   A forty-one
   B fourteen
   C forty

2. Count how many.

   A 7
   B 8
   C 9

3. Which number sentence is correct?

   A 47 < 49
   B 49 < 47
   C 47 > 49

4. Which is a fair trade?

   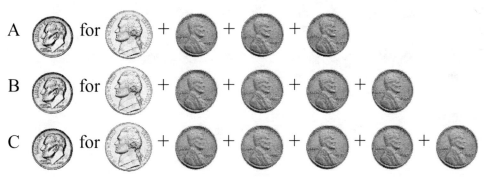

5. What is the value of the bill below?

A $20.00
B $10.00
C $1.00

M1N1f

6. Look at the numbers chart below.
   Which number is 28 closest to?

1	2	3	4	5	6	7	8	9	10
11	12	13	14	15	16	17	18	19	20
21	22	23	24	25	26	27	28	29	30

A 30
B 20
C 10

M1N2a

7. How many groups of tens?

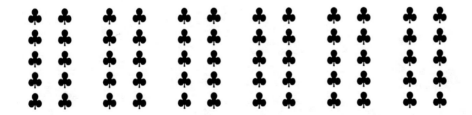

A 6 groups of tens
B 5 groups of tens
C 4 groups of tens

M1N2b

8. How many tens and ones does 53 have?

A
Tens	Ones
3	5

B
Tens	Ones
5	3

C
Tens	Ones
53	0

M1N2c

9. What is one more than 22?

   A 21
   B 222
   C 23

M1N3a

10. Which number sentence is equal to 8?

   A $3 + 5 = 8$
   B $4 + 5 = 8$
   C $5 + 5 = 8$

M1N3c

11. You have 7 acorns. You count back 2. How many acorns do you have now?

   A 7
   B 5
   C 2

M1N3e

12. Add. $\begin{array}{r}72\\+17\end{array}$

   A 79
   B 82
   C 89

13. There were 10 baby alligators in a nest.
    3 of the baby alligators stayed with the mama alligator.
    How many baby alligators have left home?

   A 7
   B 6
   C 3

14. Holly Elementary School had 15 cakes to sell at the school fair.
    They sold 10 cakes and gave the rest of the cakes to the teachers.
    How many cakes were given to the teachers?

   A 6
   B 5
   C 4

15. Glenn made 10 pieces of toast for his family of 5 people (including Glenn). Glenn gives each person the same number of slices. How many slices does each person get?

1, 2, 3, 4, 5, 6, 7, 8, 9, 10

A 5
B 3
C 2

M1N4a

16. Complete the sentence:
The triangle is cut into _____ .

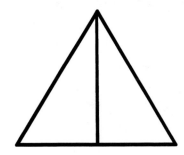

A 1 piece.
B 2 pieces.
C 3 pieces.

M1N4c

17. Which rat is the biggest?

Harry   Snuffles   Fur-Ball

A Harry
B Snuffels
C Fur-Ball

M1M1a

18. About how many paper clips long is the leaf?

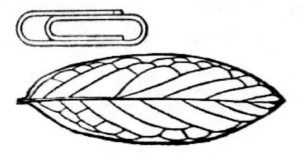

   A 1 paper clip long
   B 2 paper clips long
   C 3 paper clips long

M1M1b

19. About how many thumb widths are the glasses?
    (Use the thumb widths in the drawing. Not your own thumb width.)

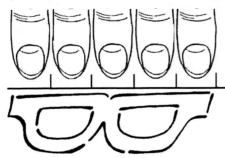

   A 3
   B 4
   C 5

M1M1c

20. What time is it to the nearest hour?

   A 8:00
   B 9:00
   C 11:00

M1M2a

21. Which takes the longest to do?

    A Write a thank you note.
    B Walk up 10 stairs.
    C Sing the alphabet song.

                                                                    M1M2c

22. What shape is most like a page from this math book?

    A square
    B rectangle
    C pentagon

                                                                    M1G1a

23. How many corners does a pentagon have?

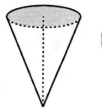

    A 3
    B 4
    C 5

                                                                    M1G1a

24. What is the name of the object that holds the ice cream?

    A cube
    B cylinder
    C cone

                                                                    M1G1b

25. What is the name of the figure below?

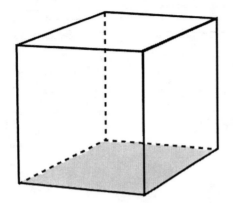

   A cube
   B cylinder
   C cone

M1G1b

26. If a shape has four equal sides, it is a _____ .

   A triangle
   B square
   C hexagon

M1G2

27. Where is the opossum?

Opossum

   A The opossum is on top of the branch.
   B The opossum is under the branch.
   C The opossum is under the ground.

M1G3

Use this chart for the next 3 problems.

Least Favorite Vegetable

	Cabbage	Onion	Radish
	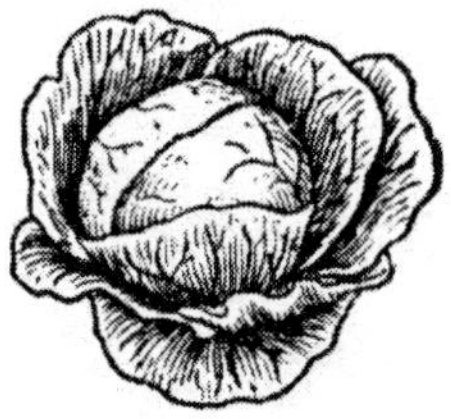	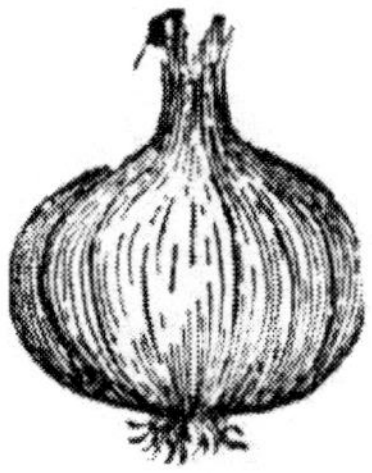	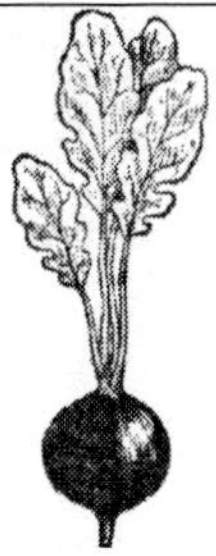
Boys	1	4	5
Girls	2	4	3

28. How many boys do not like onions?

A 5
B 4
C 1

M1D1a

29. How many girls do not like cabbage?

A 4
B 3
C 2

M1D1a

30. Which tally chart shows the vegetables the girls like least?

Chart 1	Count
Cabbage	//
Onion	///
Radish	////

Chart 2	Count
Cabbage	//
Onion	////
Radish	///

Chart 3	Count
Cabbage	///
Onion	///
Radish	////

A Chart 1
B Chart 2
C Chart 3

M1D1b

Part 2

31. The number 82 is written _____.

    A eighteen
    B eighty
    C eighty-two

32. Count how many.

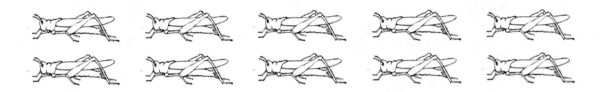

    A 8
    B 10
    C 12

33. Which number sentence is correct?

    A $94 > 93$
    B $94 < 93$
    C $93 > 94$

34. Look at the number line. What number is missing?

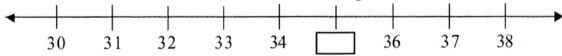

    A 33
    B 34
    C 35

35. Jack has 99 ¢. The toy he wants costs 89 ¢.
    Does Jack have enough money? Will he have money leftover?

   A No, he doesn't have enough money.
   B Yes, and he will have 10 ¢ leftover.
   C Yes, but he will have no money leftover.

36. Look at the number below.
    Which number is it closest to?

    32

    A 50
    B 40
    C 30

37. How many tens and ones does 76 have?

   A | Tens | Ones |
     |------|------|
     | 6 | 7 |

   B | Tens | Ones |
     |------|------|
     | 7 | 6 |

   C | Tens | Ones |
     |------|------|
     | 76 | 0 |

38. Skip count by 10's. What is the missing number?

    20, 30, ___, 50, 60

    A 40
    B 70
    C 80

39. Which number sentence is equal to 10?

   A 2 + 7 = 10
   B 3 + 8 = 10
   C 2 + 8 = 10

M1N3c

40. Count how many bees are in the 1st set.
    Count how many bees are in the 2nd set.
    What is Set 1 minus Set 2?

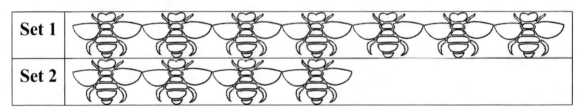

   A 2
   B 3
   C 5

M1N3d

41. Start at 37. Count backwards 2.
    What is the new number?

   A 35
   B 34
   C 33

M1N3e

42. Which sentence is true?

   A 9 + 7 = 7 + 7
   B 9 + 7 = 7 + 9
   C 9 + 7 = 2

M1N3f

43. $\begin{array}{r} 50 \\ -40 \\ \hline \end{array}$

    A 10
    B 90
    C 20

44. An owl ate 4 mice one night.
    The next night, the owl ate 2 more mice.
    How many mice did the owl eat during the two nights?

    A 2
    B 4
    C 6

45. Mrs. Jackson's chicken laid 6 eggs in one week.
    She shared them equally with 2 friends, Mrs. Smith and Mrs. Griggs.
    How many eggs will Mrs. Jackson, Mrs. Smith, and Mrs. Griggs each have?

    A 2
    B 3
    C 6

46. Is the number 51 even or odd?

    A even
    B odd
    C both

47. How is the pentagon divided?

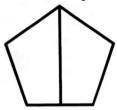

    A fourths
    B halves
    C whole

48. Which weighs the most? A car, a bicycle, or a skateboard?

    A car
    B bicycle
    C skateboard

49. About how many paperclips long is the key?

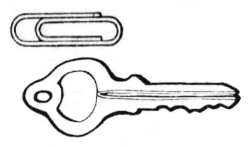

    A 1 paperclip long
    B 2 paperclips long
    C 3 paperclips long

50. What order are the days of the week?

| A | Sunday | Thursday | Tuesday | Wednesday | Monday | Friday | Saturday |

| B | Sunday | Monday | Wednesday | Tuesday | Thursday | Friday | Saturday |

| C | Sunday | Monday | Tuesday | Wednesday | Thursday | Friday | Saturday |

M1M2b

51. Which takes the <u>shortest</u> amount of time to do?

   A clap your hands 15 times

   B eat your lunch

   C play one hour at the park

M1M2c

52. Which shape is the gift card most like?

   A square
   B rectangle
   C triangle

M1G1a

53. How many corners does a hexagon have?

A 6
B 4
C 2

M1G1a

54. What are the shapes of the cylinder's faces?

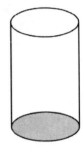

A squares
B circles
C rectangles

M1G1b

55. Which solid is the box most like?

A cube
B cylinder
C cone

M1G1b

56. Which of these shapes has 4 sides and 4 corners?

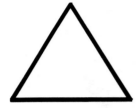  

   A triangle
   B rectangle
   C hexagon

M1G2

57. Where are the doors?

   A The doors are in the back of the barn.

   B The doors are in the front of the barn.

   C The doors are on top of the barn.

M1G3

Use this chart for the next 3 problems.

Favorite Family Pet

	Cat	Dog	Hamster
Boys	3	6	2
Girls	5	3	4

58. How many girls have cats?

    A 3

    B 4

    C 5

MlDla

59. How many boys have hamsters?

    A 2

    B 3

    C 6

MlDla

60. Which tally chart shows the pets boys have?

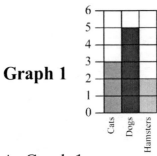
Graph 1

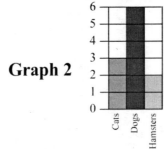
Graph 2

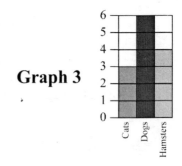
Graph 3

A Graph 1

B Graph 2

C Graph 3

MlDlb